Facilities Management Models, Methods and Tools

This book presents research tested models, methods and tools that can make the work of the facilities manager more robust and sustainable, help long-term strategic planning and support students and practitioners in FM to improve the way they approach and deal with challenges in practice.

The 34 models, methods and tools are presented in relation to five typical challenges for facilities managers:

- Strategy development
- Organisational design
- Space planning
- Building projects
- Optimisation

The chapters are short and concise, presenting a central illustration of one model, method or tool with explanatory text and short, exemplary case studies. Each chapter includes references for further reading, and the book includes a keyword index. Essential reading for all involved in the management of built assets, this book bridges the gap between robust academic research and practical industry tools. It can also be used as a handy student reference.

Per Anker Jensen is professor in Facilities Management and head of the externally funded Centre for Facilities Management – Realdania Research at the Technical University of Denmark since 2008. He holds an MSc in civil engineering and a PhD and MBA. Besides research and teaching, he has 20 years of experience as a consultant, project manager and facilities manager.

Facilities Management Models, Methods and Tools

Research Results for Practice

Edited by Per Anker Jensen

Routledge
Taylor & Francis Group

LONDON AND NEW YORK

First published 2019
by Routledge
2 Park Square, Milton Park, Abingdon, Oxon OX14 4RN

and by Routledge
52 Vanderbilt Avenue, New York, NY 10017

Routledge is an imprint of the Taylor & Francis Group, an informa business

First issued in paperback 2021

The book is based on research undertaken at the Danish national Centre for Facilities Management – Realdania Research (CFM) during the period 2008–2018. CFM has received financial support from the Danish private foundation Realdania and is administratively based at DTU Management, Technical University of Denmark. The book is based on two anniversary publications published by CFM in Danish in 2018, see www.cfm.dtu.dk.

Translations to English: The editor and other contributors
Graphics: The contributors and Hedda Bank Grafisk Design

British Library Cataloguing-in-Publication Data
A catalogue record for this book is available from the British Library

Library of Congress Cataloging-in-Publication Data
Names: Jensen, Per Anker, author.
Title: Facilities management models, methods and tools :
   research results for practice / Per Anker Jensen.
Description: Abingdon, Oxon ; New York, NY : Routledge, 2019. |
   Includes index.
Identifiers: LCCN 2019004895 | ISBN 9780367028725 (hbk) |
   ISBN 9780429001055 (ebk)
Subjects: LCSH: Facility management.
Classification: LCC TS155 .J46 2019 | DDC 658.2—dc23
LC record available at https://lccn.loc.gov/2019004895

ISBN: 978-0-367-02872-5 (hbk)
ISBN: 978-1-03-209238-6 (pbk)
ISBN: 978-0-429-00105-5 (ebk)

Typeset in Goudy
by Apex CoVantage, LLC

Contents

A.III
Space planning
37

A.IV
Building project
49

A.V
Optimisation
63

PART B
Models, methods and tools
81

B.I
Facilities that support users and activities
87

Preface

This book presents ten years of research at the Centre for Facilities Management – Realdania Research (CFM). CFM was established as a national research centre in Denmark with financial support from the private foundation Realdania on 1 January 2008 with the purpose of strengthening the research within Facilities Management (FM).

CFM has had a physical base with management, etc., at the Technical University of Denmark (DTU) and a place at the institute DTU Management, but it has also from the start been a virtual centre with researchers and collaboration partners placed at other Danish universities and companies: Roskilde University, SBi (Danish Building Research Institute/Aalborg University), University of Southern Denmark and the consulting company COWI A/S. CFM initially was established for a period of five years, but it has since been prolonged, including additional financial support from Realdania.

This means that CFM had its ten-year anniversary by the start of 2018. This was celebrated by a whole-day conference on 27 February 2018 at DTU. On that occasion, CFM launched two new publications in Danish – an anniversary book and a *Guide to Facilities Management Tools*. The present book in English includes a combination of the two anniversary publications.

The idea behind the book is to give a general overview of CFM's research in the period of 2008–2017 by presenting the most important results of practical, applicable character from CFM's research projects. It is aimed mainly at FM practitioners, but also students and others with an interest in FM.

The book is divided into two main parts. Part A is a guide to FM tools and introduces a short and concise description of 22 tools related to 5 typical processes in FM practice. Part B is a more in-depth, but still fairly short and concise, description of 34 models, methods and tools of practical application in FM.

The guide is an appetiser and entry to select relevant tools, which can then be understood deeper by reading the related chapter about a specific tool in Part B. Those really interested can find further information in the references included in each chapter in Part B.

The book as a whole has been edited by Per Anker Jensen, who has been head of CFM since the start. The guide to FM tools were, in the original Danish versions, edited by a small task force consisting of Per Anker Jensen; Susanne Balslev

Nielsen, former deputy head of CFM; and Mette Ullersted, graphical facilitator with her own company, Playmaker. It was developed from the involvement of a number of experts in research and practice in two workshops. The current English version in Part A was edited and translated by Per Anker Jensen.

Part B was also edited by Per Anker Jensen. The chapters were prepared in a collaboration between the editor and the author of each chapter, and the authors have approved the chapters based on their research. Each chapter has been supplemented by an evaluation of the practical applicability of the results, and these evaluations have been made by the editor – in some cases together with the researcher in question.

Finally, I wish to thank all the contributors to the book for their participation, all researchers and collaboration partners who have participated in CFM's activities over the ten years and not least, the foundation Realdania, which, through its financial support, has made it possible for CFM to strengthen the research in FM.

Per Anker Jensen
Head of CFM

Forewords

It is my pleasure to provide a foreword to this publication. While I am perhaps an unusual choice to be asked to do so, given my bias towards doing it and communicating it rather than theorising about it, I am, however, passionate about True Facilities Management and welcome every opportunity to support members of my peer group, who try and communicate the potential that it offers. I applaud Per Anker Jensen and his team within the CFM of Technical University of Denmark for the work that they have done over many years.

The educational and research communities that I have known over many years play important roles in the recognition and development of what is the 'new kid on the block' of the professional disciplines within the built environment sector and is indeed 'business critical' for our ongoing development as a strategic professional discipline. Per Anker Jensen is someone I have known since his entry into the educational sector of the Facilities Management community, and I have shared his vision to inform and develop awareness regarding the value that Facilities Management offers the public sector, industry and commerce worldwide.

The subject matter of this publication is important and represents a considerable amount of effort over a prolonged period of time. It represents the developing years of Facilities Management as a distinct discipline as opposed to the more common perception that it is merely a facilities services function. Therefore, it is an important record of that period of time and the evolution of the thinking, as well as the maturing of the discipline itself.

I have personally held the view that within Europe, which has led much of the development of Facilities Management as a strategic professional discipline, it has been the Nordic countries, along with perhaps the UK and the Netherlands, that have been at the forefront of the thinking within the sector. Therefore, it is in this context that I believe this publication is a significant contributor towards recording the journey that has been the evolution of Facilities Management to date and also a record of the significant contribution that has come from Denmark.

Regarding the content of the research profiled in the book, within Part A: Guide to FM Tools, you will find subject matters such as strategy development, organisational design and optimisation. All of these are fundamental to the delivery of an effective Facilities Management regime and coincidentally all are also fundamental within the recently published ISO Management System Standard for Facilities Management ISO 41001, in which I have been involved.

Within Part B: Models, Methods and Tools, my attention was drawn towards Section V titled 'FM and Added Value'. Why? Because if this is not fundamentally relevant to what Facilities Management is, then we do not have any real relevance in the organisations in which we operate!

For me, True Facilities Management is no more and no less than good management and common sense. However, I also believe that True Facilities Management is a people business. Everything that we do is ultimately about people and how we can influence them in the widest sense. Publications such as this are extremely important when considering that some of those people will be tomorrow's facilities managers and leaders. They need to have an understanding of where Facilities Management has come from and why and how it has been used to great effect across the world, and perhaps more importantly how it is being used today and harnessing technology to improve the quality of life for all as is stated in the ISO 41011 published definition, which states that Facilities Management is 'organizational function which integrates people, place and process within the built environment with the purpose of improving the quality of life of people and the productivity of the core business'.

I, therefore, applaud the work of CFM and those similar institutions around the world who undertake such research and capture it for the education and awareness of all. I especially applaud those who undertake the task of communicating it in order that others can benefit. I do not envy them and therefore will continue to focus on the 'doing' aspect of it for as long as I can.

Stan Mitchell
CEO Key Facilities Management International
Chairman ISO TC 267 Facility Management

When I was first asked to write a foreword for *Facilities Management Models, Methods and Tools*, I was slightly surprised and wondered what I could say about the publication and to others involved with the many aspects of Facilities Management.

After reflecting on my concerns, I realised that the content spoke for itself. After ten years of research activity, the Centre for Facilities Management – Realdania Research (CFM) is well placed to offer up a wealth of material, which will benefit and give support to Facilities Management professionals. These professionals include those working in education and research, as well as those who face customers and clients.

The *Facilities Management Models, Methods and Tools* is split into two distinct areas and acts as a guide to FM tools and an outline of models, methods and tools. The collection of chapters in this book provide a great deal of thought on leadership and practical suggestions. The chapters highlight some approaches, which will enable all those interested in Facilities Management to become aware of the ideas and themes that we all face to differing degrees on a regular if not daily basis.

The themes covered by the research strands range from strategic thinking to space planning to service and added value. When looking at the work that had been undertaken and the models and methods that are presented, there is a wealth of information that covers the workplace, sustainability and innovation, as well as the more recent work concerning the transfer of knowledge between Facilities Management and construction projects.

This is a publication that has much to offer to those starting out in the world of Facilities Management, but it is also a valuable resource for seasoned professionals. The content will offer a degree of new insights as well as confirm that you are on the right track with what you are doing.

Gordon Campbell McMillan
Facilities Manager (UK and Ireland) Medtronic Limited
Former Board Member of the European Facility Management Network (EuroFM)
Former Chairman of the EuroFM Corporate Associates Network Group

Facilities Management is still a young discipline in practice as well as in an academic perspective. My own way into Facilities Management started in the early 1980s from working with property and building management, and then actively joining the EuroFM and International Facility Management Association (IFMA) networks back in early 1990. In relation to these network activities, more than 20 years ago, I met Per Anker Jensen. From the first time we meet, I got to know Per as a very knowledgeable professional in Facilities Management, focussing both on the professional development in practice and developing education and research.

From 2005 when Per Anker Jensen started working on developing FM education at DTU, our collaboration has been focussed on developing education and research in FM in close cooperation with practice. This work represents a common understanding of a knowledge triangle between FM research, practice and education, where students, educators and researchers cooperate with professionals. The learning process is based on interaction, giving the possibilities for implementing research results and testing new solutions for tomorrow's FM activities. In order to succeed, the knowledge development with research and implementation of new FM models, tools and methods needs to be documented and published.

Here the new book 'Facilities Management Models, Methods and Tools' edited by Per Anker Jensen, represents an important international publication contributing to the development of FM. The new book represents the most comprehensive overview of multiple facilities management models, methods and tools collected, systemised and documented in one publication.

The overview of the research results for practice developed by CFM over a period of ten years show that FM has developed to be a recognised professional and academic field. The book describes FM in line with the European FM standard EN 15221–1 from 2006 and the new global ISO 41011 from 2017, and not at least the understanding of FM that has been the foundation of the Danish Facilities Management Association (DFM).

The previous publications 'CFM's Anniversary Book' and 'Guide to FM Tools', published in Danish, have been important for FM practitioners and education in Denmark, and useful for readers in Norway and Sweden. Taking into account that many potential readers are not familiar with the Scandinavian languages, I am glad that Per Anker Jensen has edited and translated the two books into one book published internationally in English to reach a far larger group of international readers from practice and academia.

The new book *Facilities Management Models, Methods and Tools: Research Results for Practice* is first and foremost an impressive comprehensive and thorough documentation of ten years of research at the Centre for Facilities Management (CFM) at DTU. Realdania's support for the establishment of CFM at DTU in 2008 has been central to strengthening the research and development of the FM field.

I am very pleased with the work Per Anker Jensen has done to document the CFM research, and I am looking forward to using the new book as a basis for future research and cooperation with FM practice. In addition, I am curious to see how we can use and develop the many models, methods and tools in our FM educational programmes in the coming years.

Tore Brandstveit Haugen
Professor, dr.ing. Center for Real Estate and Facilities Management
Faculty of Architecture and Planning,
NTNU – Norwegian University of Science and Technology Trondheim, Norway

Abbreviations

BIM	Building Information Model/Modelling
BREEAM	Building Research Establishment's Environmental Assessment Method
CAD	Computer Aided Design
CAFM	Computer Aided Facility Management
CAS	Campus Service at DTU
CEO	Chief Executive Officer
CFM	Centre for Facilities Management, Technical University of Denmark
CREM	Corporate Real Estate Management
CSR	Corporate Social Responsibility
DFM	Danish Facilities Management association
DGNB	German Building Sustainability Certification (Deutsche Gesellschaft für Nachhaltiges Bauen)
DTU	Technical University of Denmark (Danmarks Tekniske Universitet)
EFMC	European Facility Management Conference
EMS	Energy Management System
ESCO	Energy Service Company
EuroFM	European FM association
FM	Facilities/Facility Management
FS	Facilities/Facility Service
GIS	Geographic Information System
ICT	Information and Communication Technology
IEQ	Indoor Environmental Quality
IFMA	International Facility Management Association – based in the US
IS	Information System
IT	Information Technology
IWMS	Integrated Workplace Management System
KPI	Key Performance Indicator
LCC	Life-cycle costing
LEED	American Building Sustainability Certification (Leadership in Energy and Environmental Development)
NTNU	Norwegian University of Science and Technology (Norges Teknisk-Naturvitenskapelige Universitet)

NWoW	New Ways of Working
O&M	Operation and Maintenance
PDCA	Plan, Do, Check and Act
POE	Post Occupancy Evaluation
PPP	Public-Private Partnership
RUC	Roskilde University
SBi	The Danish Building Research Institute (Statens Byggeforsknings-institut), part of Aalborg University
SCM	Supply Chain Management
SDU	University of Southern Denmark (Syddansk Universitet)
SFM	Sustainable FM
SLA	Service Level Agreement
SWOT	Strengths, Weaknesses, Opportunities and Threats
VAM	Value-Adding Management

Contributors

Jakob Brinkø Berg (born Jakob Berg Johansen) earned a master of science degree in civil engineering from DTU in 2014 and is a PhD fellow at DTU Management with Per Anker Jensen and Christian Thuesen as supervisors. The project is part of the societal partnership REBUS on Renovating Buildings Sustainably supported by the Danish Innovation Fund. Chapters A.16 and B.20 are based on the PhD study. Jakob can be contacted by e-mail: jajoh@dtu.dk.

Rikke Brinkø Berg (born Rikke Brinkø) was educated as a civil engineer at DTU in 2013 with specialisation in urban and construction management, and she completed a PhD study concerning Facilities Management and Shared Space at CFM during the period of 2012–2016, with Susanne Balslev Nielsen as the main supervisor and Juriaan van Meel as the co-supervisor. Chapters A.11 and B.5 are based on her PhD study. Rikke was granted the European Researcher of the Year Award in 2016. She worked as a postdoctoral researcher at DTU Management in 2016–2017 and is now working as the space manager at DTU Campus Service. Rikke can be contacted by e-mail: rikbk@dtu.dk.

Rimante A. Cox earned a master's in environmental design from University College London in 2007. She did a PhD study on climate change and the impact on operation and maintenance of buildings at DTU, co-funded by CFM, and in collaboration with Gentofte Municipality, north of Copenhagen, in the period of 2011–2015. Carsten Rode, DTU Civil Engineering was the main supervisor and Susanne Balslev Nielsen was the co-supervisor. Chapters A.14 and B.8 are based on her PhD study. Today, Rimante lives in France and manages the property of Chateau des Egaux. Rimante can be contacted by e-mail; rimantecox@googlemail.com.

Torben Damgaard is a candidate in social science from Odense University in 1989 and has a PhD from the Southern Denmark Business School in 1997. Since 1998, he has been an associate professor at the University of Southern Denmark in Kolding. In the period of 2009–2013, he was project manager on a research project concerning implementation of operational knowledge in building projects for CFM. Chapter B.23 presents results from this project. Torben can be contacted by e-mail: torben@sam.sdu.dk.

Poul Henrik Due is educated as a geologist from Copenhagen University in 1984. Since 2012, he has been chief consultant at the consulting company Sweco. Earlier, he was employed at the Danish Technological Institute where he was managing the secretariat of the Danish Facilities Management association. During the period of 2007–2012, he worked for the consulting company COWI, and as part of that, he was project manager on a best practice project on operational-oriented building processes for CFM. Chapter B.22 presents results from this project. Poul can be contacted by e-mail: poulhenrik.due@ sweco.dk.

Poul Ebbesen is a senior specialist and project manager in the consulting company Ramboll Denmark. Over the last 20 years, he has worked with implementation and use of information technology in the Facilities Management and real estate business areas. He is responsible for projects involving analysis of real estate portfolios, development of real estate strategies and implementation of technologies. All projects involve the use of technologies such as Building Information Modelling, Computer Aided Design, Geographical Information Systems and Computer Aided Facilities Management systems. In 2013–2016, he did a PhD at DTU on value-adding digitalisation in FM. Chapters A.22 and B.28 are based on results from his PhD study. Poul can be contacted by e-mail: pou@ramboll.dk.

Aneta Fronczek-Munter received an Msc in architecture and engineering from the Technical University in Poznan, Poland, in 2004. She practised as an architect during the period of 2005–2009, including in her own company. From 2010 to 2016, she was a PhD student at CFM with Per Anker Jensen as the main supervisor and Juriaan van Meel and Werner Sperschneider as co-supervisors, and from 2016 to 2018, she was a postdoc at NTNU, Faculty of Architecture and Design in Trondheim, Norway. Aneta now works for a Norwegian agency for hospital planning and construction – Sykehusbygg HF as principal advisor for architecture and health. In 2017, she won a European Design Research Award for a conference paper based on her PhD study about usability briefing. Chapters A.13 and B.3 are also based on the PhD study. Aneta can be contacted by e-mail aneta.f.munter@gmail.com.

Kirsten Ramskov Galamba was educated as a biologist at Copenhagen University in 1999. She did a PhD study at CFM on FM and sustainable development in collaboration with Albertslund Municipality, west of Copenhagen, during the period from 2008 to 2012, with Susanne Balslev Nielsen as the main supervisor. Chapters A.7 and B.7 are based on her PhD study. Today, Kirsten works as the head of Section for Work Environment, Preparedness and Sustainability in DTU Campus Service. Kirsten can be contacted by e-mail kirg@dtu.dk.

Anne Vorre Hansen was educated as an anthropologist at Copenhagen University in 2005 and earned a PhD in service innovation from Roskilde University (RUC) in 2016. Ada Scupola, RUC, was the main supervisor and Per Anker Jensen, CFM, was the co-supervisor. Chapters A.20 and B.17 are based on her

PhD study. Today, she is a postdoc at RUC and an independent consultant. Anne can be contacted by e-mail: vorre@ruc.dk.

Birgitte Hoffmann was educated as a civil engineer and earned her PhD from DTU. She was employed at DTU until 2012, and today, she works as an associate professor at Aalborg University in Copenhagen. She specialises in the redesign of cities for liveability and sustainability, focussing on the transition of planning, professional practices and participation processes. During the period of 2008–2012, Birgitte was connected to CFM as a manager of the research project Facilities for Creative Environment, presented in Chapters A.12 and B.2. Birgitte can be contacted by e-mail: bhof@plan.aau.dk.

Jesper Ole Jensen was educated as a civil engineer at DTU in 1991 and earned a PhD degree from Aalborg University in 2001. He works as a senior researcher at the Danish Building Research Institute (SBi), Aalborg University, with specialisation in buildings, city development and sustainability. During the period from 2009 to 2012, he was lead researcher on CFM's project on Energy Service Companies (ESCO) together with Susanne Balslev Nielsen, among others. Chapter B.9 is based on the project. Jesper Ole can be contacted by e-mail; joj@sbi.aau.dk.

Per Anker Jensen was educated as a civil engineer at DTU in 1978 and earned his PhD in 1985 and his MBA at Copenhagen Business School in 2014. For 20 years, he worked as a consultant, project manager, facilities manager and deputy project manager in the client organisation of a new media centre for the Danish Broadcasting Corporation. Since 2005, he has worked on developing FM as a new subject at DTU, and since 2008, he has been the head of CFM. During the period from 2009 to 2016, he was appointed a professorship with special assignments in FM. He is now the associate professor and coordinator for FM at DTU Management. Per has been and is the main supervisor and co-supervisor for many of CFM's PhD students. Many chapters in both parts A and B present results from his research. Per can be contacted by e-mail; pank@dtu.dk.

Akarapong Katchamart earned a bachelor's degree in food technology from Chulalongkorn University, Thailand, in 2005 and has a master's degree in facility management from Pratt Institute, New York, in 2009. He did a PhD study about FM and added value at CFM from 2009 to 2012, with Per Anker Jensen as the main supervisor and Juriaan van Meel as the co-supervisor. Chapters A.2 and B.26 are based on results from the PhD study. Today, he works in FM at a public hospital in Bangkok. Akarapong can be contacted by e-mail: akatchamart@gmail.com.

Kristian Kristiansen was educated as a sociologist at Copenhagen University in 1980 and has a master's degree in social science from Roskilde University. After several years as consultant, senior researcher at the Danish Building Research Institute and secretarial manager for the Danish Building Development council, he became an associate professor at DTU and head of studies of a postgraduate master programme in construction management. From 2009 to 2013, he was responsible for a research project at CFM about public-private partnerships and FM. Chapter B.18 is based on this project. Kristian can be contacted by e-mail: krk@kabelmail.dk.

Esmir Maslesa was educated as a civil engineer at DTU in 2011. Afterwards, Esmir worked for CFM from 2011 to 2014 as a research assistant on a project concerning sustainable building renovation (see Chapters A.15 and B.10). At present, Esmir is an industrial PhD fellow researching how FM systems and dynamic data can be used to optimise environmental building performance. The PhD study takes place in the period from 2017 to 2019 in collaboration with the Information Technology company KMD. Chapter B-12 is based on results from the PhD study. Esmir can be contacted by e-mail: esmas@dtu.dk.

Giulia Nardelli earned a master's degree in business management from both Bocconi University (Milan) and Copenhagen Business School in 2010, and earned a PhD from Roskilde University (RUC) in 2014 with Ada Scupola, RUC, as the main supervisor and Per Anker Jensen, CFM, as the co-supervisor. She was granted the European Researcher of the Year Award in 2014. In 2014, she worked as a postdoc at CFM, and today, she is an assistant professor at DTU Management. Chapters A.4, A.19, B.14, B.15 and B.16 are based on the results of her research, but the chapters were prepared by Per Anker Jensen. Giulia can be contacted by e-mail: ginar@dtu.dk.

Susanne Balslev Nielsen was educated as a civil engineer at DTU in 1993 with a specialisation in sustainable city development, and she did a PhD study on the transition of municipal infrastructure in 1994–1998. Until the end of 2016, Susanne worked as an associate professor at DTU, and she was the deputy head of CFM with a main focus on sustainable FM. She has been the main supervisor and co-supervisor of several of CFM's PhD students. Susanne was granted the European Researcher of the Year Award in 2010. Some of her research results are presented in Chapters A.3, A.8, B.6 and B.33. She now works as the FM expertise director at the consulting engineering company NIRAS. Susanne can be contacted by e-mail: subn@niras.dk.

Helle Lohmann Rasmussen graduated as an architect in 2003 from Aarhus School of Architecture. She later earned a master of construction management degree from DTU Management in 2014. She has worked for several years as a project manager of building-related projects, including at DTU Campus Service. Since December 2016, she has been a PhD student at CFM with Per Anker Jensen as the main supervisor and Jay Sterling Gregg as the co-supervisor. Her PhD project is about the transfer of operational knowledge to design. Chapter B.24 is based on the PhD study. Helle can be contacted by e-mail: helr@dtu.dk.

Ada Scupola received an Msc at the University of Bari, Italy, in 1986, and earned a PhD in business administration at Roskilde University, Denmark, in 2000 and an MBA at the University of Maryland, USA, in 1990. She is currently a professor at Roskilde University, and since 2008, she has been involved in collaboration with CFM – for instance, with her own research in service innovation and as a supervisor of PhD students. Chapter B.13 presents results from her research. Ada can be contacted by e-mail: ada@ruc.dk.

Christian Stenqvist was educated as a civil engineer and earned a PhD in environmental and energy system analysis at Lund University, Sweden, in 2014. In 2014, he was the manager of a development project on Energy-efficient FM, which was carried out in collaboration with the Swedish consultant company EVU Energi & VVS Utveckling AB and CFM with support from the EU's regional development fund (Interreg Øresund-Kattegat-Skagerrak). Chapters A.21 and B.11 are based on the results of this project. Today, he works as an independent consultant within programme and project evaluation within the energy area in Sweden. Christian can be contacted by e-mail: christian.stenqvist@evalpart.se.

Kresten Storgaard was educated as a sociologist at Copenhagen University in 1974. He has for many years worked as a researcher at the Danish Building Research Institute (SBi), today part of Aalborg University. In the period from 2009 to 2013, he was managing CFM's research project on long-lasting collaborations within FM. Chapters A.6 and B.19 present results from the project. Kresten can be contacted by e-mail: krestenstorgaard@gmail.com

Mette Tinsfeldt was educated as a civil engineer at DTU in 2013. She did her master's thesis with Per Anker Jensen as the supervisor. Chapters A.10 and B.4 are based on the thesis and a conference paper. Today, Mette is employed as a senior project manager in the corporate real estate department of one of the main financial institutions in Denmark, Danske Bank. Mette can be contacted by e-mail mtinsfeldt@gmail.com.

Juriaan van Meel was educated as an architect and earned a PhD at Delft University of Technology in 2000. In the period from 2009 to 2015, he was engaged at CFM as a senior researcher on a research project concerning workplace management, which is presented in Chapters A.9 and B.1. Juriaan has also been a co-supervisor of PhD projects at CFM. In addition, Juriaan is the co-founder and partner at the planning company ICOP and the software company BriefBuilder, with bases in Rotterdam and Copenhagen. Juriaan can be contacted by e-mail: vanmeel@icop.nl; ICOP: www.icop.nl.

Introduction

This book is for those of you who work with Facilities Management (FM) in practice or as students. It presents 34 models, methods and tools that can make your work and decisions more robust and sustainable, and can, for instance, help you to change from firefighting towards long-term planning or from de-centralisation to centralisation. Or help you to develop a strategy with a broad-based support from the organisation.

The book presents the results of ten years of research at the Centre for Facilities Management – Realdania Research (CFM). CFM was established as a national research centre in Denmark with financial support from the private foundation Realdania on 1 January 2008 with the purpose of strengthening the research within Facilities Management.

This means that CFM had its ten-year anniversary by the start of 2018. This was celebrated with a whole day conference on 27 February 2018 at DTU. On that occasion, CFM launched two new publications in Danish: an anniversary book and a *Guide to Facilities Management Tools*. The present book in English includes a combination of the two anniversary publications.

The idea behind the book is to give a general overview of CFM's research in the period from 2008 to 2017 by presenting the most important results of the practical applicable character from CFM's research projects. The aim is to support students and practitioners in FM to improve the way they approach and deal with challenges in practice and help them to do their work and make decisions that are more robust and sustainable. The book is divided into two main parts.

The first part is a guide to 22 new tools for FM presented in an easily accessible way. The guide is organised in relation to five typical processes, that facilities managers often are involved in during their daily work. Four or five tools are presented for each process. The FM staff intended as users of the processes and tools are identified as five-person profiles. Each tool is presented with texts describing what, why, who and how, and an illustration with explanation. There is for each tool a reference to a chapter in the second part of the book where the tool is presented in more depth.

The second part consists of 34 short chapters, which cover a project or result, including a central illustration of a model, method and/or tool, and often together with short case presentations. Twenty-two of the chapters concern tools that are introduced briefly in the first part of the book. Texts and illustrations are mostly based on earlier publications, but they have been edited in a standardised and concise form. Each chapter includes references to the most important literature with more detailed information about the results.

Part A
Guide to FM tools

Introduction to part A

The guide starts with the following five typical challenges that facilities managers meet in their daily work life:

1 Strategy development
2 Organisational design
3 Space planning
4 Building projects
5 Optimisation

For each challenge, a process is described to give you inspiration, and for each process, there are four or five tools which you and your team can make use of. This is all to help you make more long-lasting and robust decisions and develop better planning.

The term tools should be understood in a broad sense as a knowledge-based means to be used in work with FM. This can include frames for understanding, concepts, terminologies, typologies, methods, models and step-wise procedures. The criteria for selection and presentation has been that the tools have practical application, but it, of course, varies as to how easy they are to use.

The guide in Part A and the chapters in Part B of the book supplement each other, with the guide being an entry to and overview of the tools. If you want to learn more about the tools and make use of one of them, then it is a good idea to read the chapter in Part B about the tool in question. Each chapter in Part B has an overview of further literature about the specific tool, which is aimed at those who want to dig deeper into the research basis for the development of the tool.

What are the processes about?

Independent of where you are as a facilities manager and how you work, you are likely to be involved in some of these processes. Which do you have to deal with soon, and which do you need to focus on in a longer time perspective?

Strategy development: Concerns the formulation of the direction for the development of the FM activities and defines organisational goals. It forces you to raise your eyes from the daily tasks and outline the long-term perspectives.

Organisational design: The structure and culture of collaboration in the organisation is central to any head of FM. FM is a management discipline that requires holistic thinking and coordination. Therefore, it is necessary from time to time to consider whether the existing organisation can be more optimal in relation to your needs and development goals.

Space planning: This is related to space management and workplace management. It requires that you know your building portfolio, know the needs of the users and are capable of facilitating and coordinating development processes, which involves relocation, new office layout and changes in services.

Building projects: Concerns developing more sustainable buildings in a life-cycle perspective by focussing on the future users and ensuring that buildings are operation friendly by integrating knowledge from FM.

Optimisation: Concerns the FM organisation reconsidering each working procedure. This can be caused by new strategic goals or insufficient performance in measurement of – for instance, satisfaction with response time, service levels or cost levels.

Who are you?

The processes and tools are described for you, who looks like one of the following five people or have tasks and functions that resembles theirs.

Independent of whether you recognise yourself mostly in Hanna, John, Carl, Karen or Evelyn, the guide is aimed at people who

- Take or get a role as a 'process manager' and drive meetings, discussions and development
- Have responsibility for the processes in an in-house FM organisation in a large company or a new FM unit

The processes can also take place in a service provider company or a consulting company, but then the organisational context will be different.

Head of FM: Hanna is new in the position as head of FM in a larger company. She was appointed to, among other things, create an overview of the building portfolio and develop a FM strategy for the next five years. She is the leader of five section managers who all together are leading a staff of 140. Hanna is 54 years old, married and a mother of two children. Earlier, she worked in economy and administration.

FM middle manager: John is a middle manager in the FM organisation with functional responsibility for some of the buildings. The FM organisation is in the middle of a major restructuring, and John wants to succeed in giving a new direction for his staff of 20, and at the same time, he wants to ensure that all six teams will work with him and not against him. John was educated as an architect; he is 38 years old, single and the father of three.

FM team leader: Carl was initially trained as an electrician and has for a few years functioned as a team leader for approximately ten electricians in the organisation. During periods with larger projects, he also has external electricians on his team. He prepares the work plans for the team and divides the work tasks. He is occupied with developing the team and the staff's competences and mind-sets. Carl has ambitions to get more responsibility and influence, and to be promoted higher up in the FM organisation. He is 29 years and lives with his girlfriend.

Building project manager: Karen is newly employed as project manager in the building section. She worked as an assistant on building projects for a few years, but she has now become responsible for rebuilding and extension projects. She wants to succeed in developing more operation friendly buildings and has focussed on involving the operational FM staff in the building projects. Karen was educated as a building engineer. She is 30 years old and the mother of a small boy.

Project manager and developer: Evelyn is the head of development and the one in the FM organisation with continuous focus on customer relations. She is employed to drive development projects, which typically involves the FM organisation, the users and the external stakeholders. She has particular focus on coordinating the strategic ambitions of the corporate top management with the development of the middle managers' tasks and room to manoeuvre and to find practical solutions that function in daily life. She works generally with professionalising the FM organisation and sees great potential in collaboration with students and others in the industry, nationally and internationally. Evelyn is 46 years old, divorced and lives with her two teenage children.

A.1

Strategy development

Opening dialogue

Hanna, head of FM, to Evelyn, project manager and developer:

- With the new corporate strategy, we need to update our FM strategy. I am asking you to prepare a proposal for a process plan for the strategy development, which we can discuss.

Evelyn's response:

- OK. I will make a proposal for the next management meeting.

Why this process?

The process shall ensure that the FM organisation has well-defined, long-term goals and development plans which:

- Are in accordance with the strategy and objectives of the core business, and
- Makes it possible to monitor the development of the FM organisations over time.

What triggers starting this process?

- When corporate management decides to establish a dedicated FM organisation
- When major changes occur in the strategy and conditions of the core business
- When major changes occur in the tasks and the provider's situation for the FM organisation

The strategy must be evaluated and reconsidered annually or with a few years' interval – for instance, when the current strategy and development plans have been fulfilled or have become outdated.

The process

The process for strategy development consists of six phases, which are described in Table A.1.

Table A.1 Phases in the process for strategy development

Phase	Name	Description
A	**Understand corporate strategy**	It is essential that the strategy for the FM organisation is based on an in-depth understanding of the core business with regard to, among other things, its mission, vision, values, strategies, organisation and production processes so that the FM strategy supports the long-term goals and development needs of the corporation. Hints for working with phase A: 1 Search for mission, vision and strategy papers, and for central terms and focus areas which you can relate directly to FM. 2 Find and translate the corporate values to physical frames. 3 Consider whether your corporation and production processes have special characteristics which should influence the FM strategy.
B	**Where are we now?**	The FM strategy should also be based on a thorough analysis of the current situation for FM – for instance, by a SWOT analysis of current strengths and weaknesses, and external opportunities and threats.
C	**Define strategy goals**	Based on the analysis of the current situation and an investigation of the relevant directions for development – for instance, by collecting knowledge and experiences from professional literature, experts and other companies – the long-term goals for the development of the FM organisation are defined. Examples of strategy goals: • The property portfolio in the home country should be increased by 20% over the next ten years, mainly by more regional offices • In other countries, the number of rented sales offices should be increased by 30%. • Preventive maintenance of buildings should be increased compared to acute maintenance with the same budget per square metre. • The amount of flexible offices should be increased to 75% in five years. • The canteens should be changed to serve more healthy food with a minimum of 90% being organic.
D	**Develop strategy plans**	Based on the strategic goals, plans are prepared for how to achieve the goals, and this is specified in a programme of actions and an investment plan divided in a number of initiatives and projects.

Phase	Name	Description
E	Implementation and follow-up	Based on the programme of actions, the individual initiatives and projects are initiated in a prescribed order, and persons responsible are appointed for each initiative and project. The management of the FM organisation is monitoring the implementation by continuous follow-up based on status reports and initiates necessary adjustments.
F	Evaluate plans	By regular intervals – for instance, once a year – the management of the FM organisation evaluates and reconsiders the FM strategy, including the programme of actions and the investment plan, and decides changes in the overall strategy and the plans. If needed, a new strategy process is initiated.

Tools for strategy development

Figure A.1 shows the names of four tools that can be used for strategy developments in FM and the phases in the process where they are most relevant. A short description of each tool is given next, and this is followed by a section about each tool.

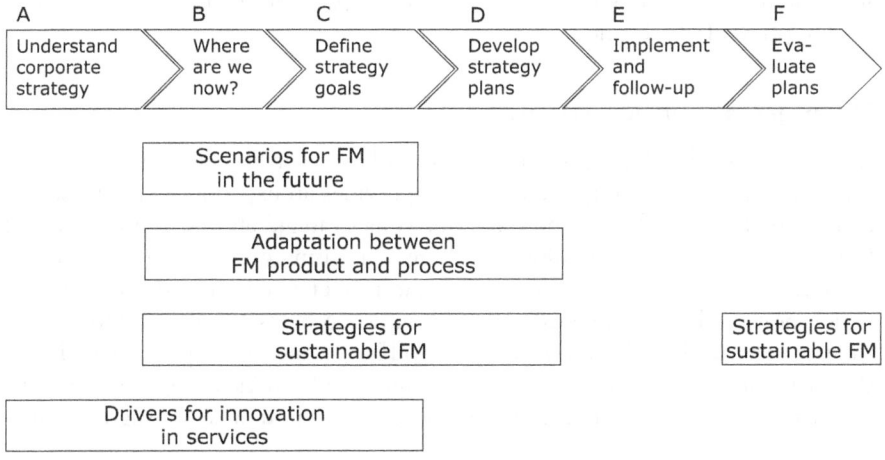

Figure A.1 Tools for strategy development and the process phases

Scenarios for FM in the future

The tool is applicable when you need to take the real broad perspective: How will international megatrends influence our FM, and how shall we strategically meet the changes? The tool provides you with a system to evaluate the possible impact of major global trends on your organisation and FM strategy. It can, in particular, be useful in phases B and C, where you evaluate your present situation and define your strategic goal.

Adaptations between FM product and process

The tool provides you with an overview of where you are now: whether you have chosen to make partnerships with external providers or you offer your provision in-house. The model gives you, management of the FM organisation and top management, a common understanding of how you have chosen to organise the FM activities strategically and implement FM in your company. It is a good starting point for making decisions about how you want to develop your FM organisation, and can be used in phases B, C and D.

Strategies for sustainable FM

There are many approaches to work with sustainability, and therefore it is important that you make a conscious choice about, which approach you choose to take. This tool shows a number of different approaches, each with its own strengths and weaknesses. The question is, which approach is most suitable for your particular organisation and your motivation to work with sustainability. The tool can specifically give you and your top management a shared picture of the strategy for sustainability you follow today and a frame for your future ambitions. It can be used in phases B, C, D and F.

Drivers for innovation in services

If your strategy development concerns rethinking your FM services because the surroundings are changing, then this tool provides an explanatory model which can be used to initially set the frame for your strategic efforts. You can benefit from using the tool to verbalise that there is a mismatch between your current services, which were customised to specific former needs, and the new needs, which now are being expressed. The innovation process consists of understanding the new needs and translating them to FM services, which are aligned with the other strategic values, economical frames and practical applicability. The new services can then, if necessary, be implemented in new service agreements.

A.1 Scenarios for FM in the future

What?

The tool is a procedure and a model that helps you to identify future trends, which can influence your organisation and your FM strategy.

Figure A1.1 shows five scenarios named with traffic metaphors for FM in the future. These were identified based on alternative development in two major societal frame conditions: The general economic development in Northern Europe and the global climate.

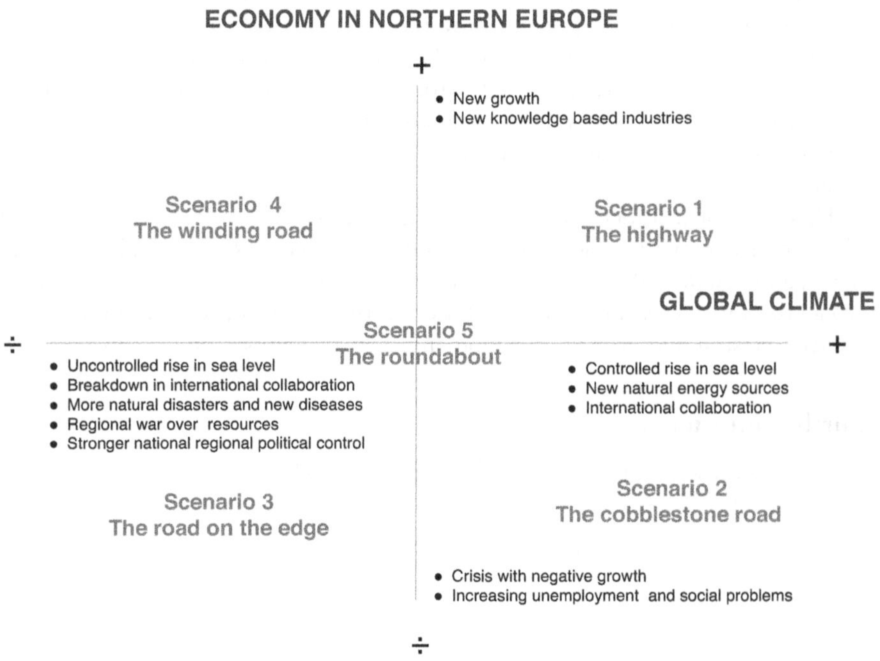

ECONOMY IN NORTHERN EUROPE

+

- New growth
- New knowledge based industries

Scenario 4
The winding road

Scenario 1
The highway

GLOBAL CLIMATE

Scenario 5
The roundabout

÷   **+**

- Uncontrolled rise in sea level
- Breakdown in international collaboration
- More natural disasters and new diseases
- Regional war over resources
- Stronger national regional political control

- Controlled rise in sea level
- New natural energy sources
- International collaboration

Scenario 3
The road on the edge

Scenario 2
The cobblestone road

- Crisis with negative growth
- Increasing unemployment and social problems

÷

Figure A1.1 Analytical model with examples of scenarios

Why?

When you use the tool, you will get inspiration and a systematic evaluation of which impact selected global megatrends can have on your organisation and FM strategy. It provides a visual overview, and when you have mapped trends and possible consequences and discussed external and internal influences, you will avoid sitting in an ivory tower to define your long-term strategic goals.

Who?

The tool is used by FM management and possibly with involvement of internal key persons and external specialists. This typically takes place during a number of workshops with related preparations and follow-up processing and reporting.

How?

You initially conduct a brainstorm about which general trends in society can have major impact for you in the future. It is a good idea to beforehand read various reports with future studies and perhaps get inspirational presentations from external experts. As facilitator, you have as preparation studied the tool so that you can present the model as a frame for understanding and an analytical tool for the others in FM management group. You should also have considered which relevant persons and experts can supplement your own perspective.

The megatrends which you expect to have the greatest importance and which at the same time are uncertain should be discussed further and investigated, for instance, by the use of literature on the topic and involvement of external experts. You decide on two trends as the main factors and at a new workshop, you develop scenarios for what combinations of these two factors will possibly influence your organisation.

This can, for instance, be by combining a best-case, a worst-case and a most likely scenario for each factor. Each combined scenario is given a characteristic name, and you describe it as detailed as possible – for instance, divided in smaller groups during or after the workshop. Based on the scenarios, you can then develop alternative strategies for the development of your organisation.

Further information

The tool is presented in more detail in Chapter B.34.

A.2 Adaptation between FM product and process

What?

The model contributes to clarify what characterises the different FM provisions – in the model called products, which an organisation demands, and how the process for delivering the provisions is arranged in the most appropriate way. The products and the process make up the two axes in a matrix. The need for FM provisions and the relation between the FM organisation and the customers can be in different positions in the matrix. For products, there is a differentiation between the degree of complexity and the need for customisation, and for the process, there is a differentiation between the degree of interaction between providers and customers.

The model points to products with low complexity, and customisation are best delivered with an arm's-length relationship between the customer and the provider. A higher degree of complexity and customisation will on the other hand favour establishing a partnership, which can be based on a transaction relation or – in case of even higher complexity and customisation – an operational partnership. With the highest degree of complexity and customisation, a vertical integration can be the right solution, for instance, with in-house provision, establishing a subsidiary company and establishing a shared company between customer and provider or by a strategic partnership. Different FM services often have different characteristics, thus a combination of different delivery processes to an organisation can be relevant.

Figure A2.1 shows different strategic positions from basic, standardised FM services to a preferred partner that delivers bespoke services.

Why?

You can use the tool when you need to discuss with top management how you strategically have implemented – or should implement – FM in the corporation. The model provides a terminology and conceptual framework that can help you to analyse which types of collaboration are suitable for different FM services. This is relevant when you, for instance, define goals or evaluate current service levels and forms of collaboration in the FM organisation.

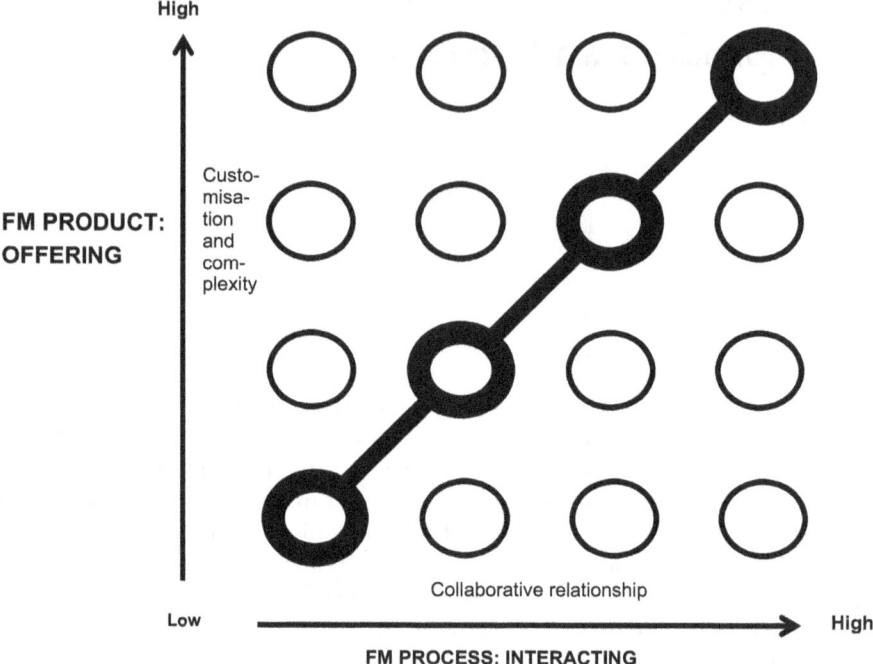

Figure A2.1 Product/process matrix for FM

When you are finished, you have a common picture of what characterises the organisation's demand for FM provisions and a basis to discuss whether the delivery process is adapted in an optimal way in relation to the demand or whether there is a reason to change the way you have arranged your FM provisions.

Who?

The head of FM and top management of the corporation together decide how the FM provisions are delivered in the best possible way. This typically takes place over one or more meetings, which focus on the general product/process strategy and decisions are made concerning each of the areas of the provisions from the FM organisation.

How?

Describe the current situation of how FM is implemented in the corporation today, and create an overview of your FM products, as well as the way you collaborate with providers of the different products. Discuss, based on the model in Figure A2.1, the current situation to establish a common understanding. Then focus on the preferred development: How should the different FM products and the collaborative relationships between the core business, the internal FM organisation and the external providers be arranged in the future?

OBS!

Beware talking about new ideas and possible changes too soon. It is important to create a common understanding of the present situation and the advantages and disadvantages there are with the way the delivery of FM products is arranged today.

Related tools

The tool can be used together with the *organisation of FM in relation to core business* (see Chapter A.5) and *collaboration with external providers* (see Chapter A.6).

Further information

The tool is presented in more detail in Chapter B.26.

A.3 Strategies for sustainable FM

What?

The tool can be used to set some corner markers for the development work, which is part of integrating sustainability in FM:

- Who is the strategy related to (internal or also external partners)?
- What is the time horizon for the strategy? (Is it the present situation and expectation about a known future or are more fundamental society changes anticipated?)

The model includes three sustainability strategies for FM with different ambitions:

- The ecological building: ambitions concerning building projects and rebuilding
- Environmentally friendly FM: ambitions concerning environmental resource consumption and climate impact
- Sustainable society: ambitions for the industry and societal development

The general strategies can be used individually or in combinations.

Figure A3.1 shows four generic strategic positions for sustainability and three sustainability strategies for FM related to them.

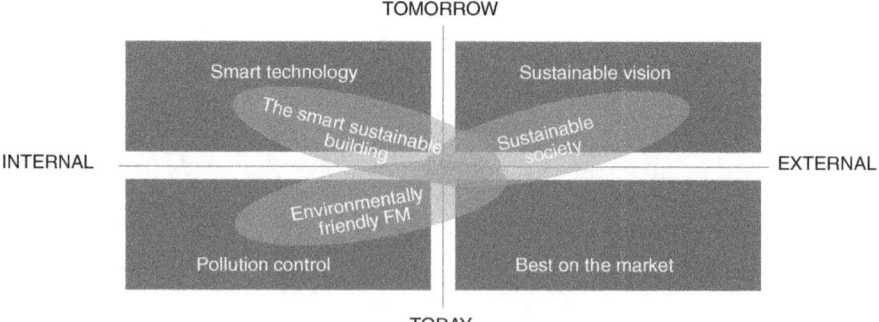

Figure A3.1 Model for sustainability strategies

Why?

The tool helps top management to imagine different sustainability strategies better. Furthermore, it helps you to 'get into the brains' of top management and understand their logic so that together you can formulate visions and goals, and you can also take action on your own. The tool provides you with a common understanding of the sustainability strategy that the organisation follows today and not only a shared understanding that is just 'on paper'. You get a visual model, which clearly shows your strategic choices and rejects in relation to sustainability which you can use to communicate further in the FM organisation.

Who?

The tool can be used by the sustainability expert in the FM organisation to develop the strategy for sustainable FM together with the management of FM. The tool is used together with your proposals for potential strategies in a dialogue with top management with the aim of deciding on the strategy of sustainability in relation to FM in the company.

How?

You should as an expert in sustainability know the three different strategies and their positions, and use them to develop proposals for your own potential sustainability strategy for FM. At a meeting with top management, you can use the model to explain first about the two axes and the four positions, and next about the current strategy for sustainability as you see it.

Consider whether your goal is to further understand the logic of top management, to try to change their attitude or to give them an understanding of possible differences between what the company says it does and what it actually does in relation to sustainability.

Afterwards, you can prepare a full strategy document which describes your ambitions, the actions needed to fulfil those ambitions and how you will know whether you succeed.

OBS!

Beware talking about new ideas and possible changes too soon. It is important to create a common understanding of the present situation, and the advantages and disadvantages there are now, before you decide on changes.

Related tools

The tool can be used together with *climate change and buildings* (see Chapter A.14) and *sustainable building renovation* (see Chapter A.15).

Further information

The tool is presented in more detail in Chapter B.6.

A.4 Drivers for innovation in services

What?

The model takes as a starting point that changes often are caused by critical incidents, which can be externally induced as, for instance, the financial crisis or internally induced as, for instance, decisions about organisational changes (see the case in Chapter B.14). The model shows that such impacts create an imbalance between needs and expectations among different stakeholders, which can lead to a conflict between the parties. The conflict must be solved by making changes in FM services, provisions and/or collaborative relationships. In that way, one can establish a new balance with the consciousness that it can be temporary.

Figure A4.1 shows the dialectic process model that explains drivers for innovation in services.

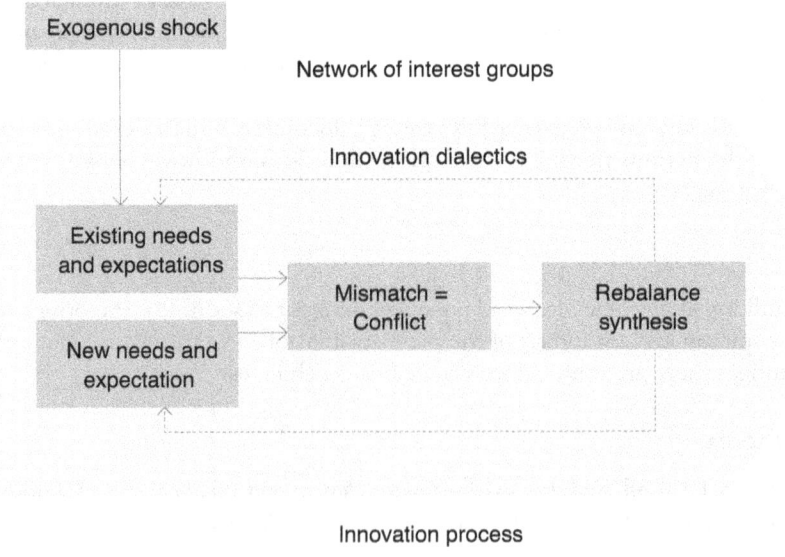

Figure A4.1 Dialectic process model for innovation in services

Why?

The model is an analytical tool that can help you identify the causes for the following:

- Changes in the company's need for FM
- What is required to change the way the provision of FM is arranged

The model can also be used to identify which stakeholders will mostly be influenced by the changes. You can use the model in the initial phases in the strategy process to understand the corporate strategy and the need to change the FM strategy and develop new strategic plans. With the model, you and your staff/colleagues on the strategic level can obtain a broader understanding of FM changing over time and what causes such changes. Based on this, you can work more systematically with innovation and develop more robust FM strategies and development plans.

Who?

Managers and staff on the strategic level in the FM organisation work with the model over a longer period in connection with strategy development. The model can also be used in relation to individual FM services.

How?

Before you start, it is necessary to have an overview of how and why the corporate strategy has developed and what the background is for the current FM strategy. You should also consider what a sufficient basis for decisions is – for instance, which data and information you need to have. Do you have a sufficient knowledge base, or do you need to collect and analyse data and information?

The model is best used as an integrated part of the initial phases of the strategy process as an analytical tool, including analysing the needs and expectations of the different stakeholders. The model is suitable as a basis for discussions among those involved in the strategy process as a common frame of understanding. Therefore, it is advantageous that you have drawn and explained it so that it is visual and can be referred to during your meetings.

Related tools

The tool can be used together with *collaboration on value creation in FM innovation* (see Chapter A.19) and *Personas as a basis for service innovation* (see Chapter A.20).

Further information

The tool is presented in more detail in Chapter B.14.

A.II

Organisational design

Opening dialogue

John, FM middle manager, to Hanna, head of FM:

- Here is as agreed a proposal for an organisational plan for a team to be responsible for FM support in the company's new production facilities in Slovakia.

Hanna's response:

- Thanks, John. I look forward to reading that.

Why this process?

The organisational design process shall ensure that the FM organisation is suitable in relation to the following:

- The tasks you are responsible for
- The competences that are involved
- The collaborative relationships the organisation has internally in the company and with external providers and collaboration partners

What triggers starting this process?

- When corporate management decides to establish a dedicated FM organisation
- When major changes occur in the strategy and conditions of the core business
- When major changes occur in the tasks and the provider situation for the FM organisation

The organisational design is evaluated and reconsidered with some years' interval – for instance, when there are changes in the strategy of the FM organisation.

The process

The process for organisational design consists of six phases, which are described in Table A.2.

Table A.2 Phases in the process for organisational design

Phase	Name	Description
A	**Purpose of changing**	As a basis for starting the process, the purpose of developing a new organisational design should be clarified between the corporate top management and the manager of the FM organisation. Hints when defining purpose: <ul><li>It is a good idea, along with the purpose, to define some specific success criteria so that you can measure whether the purpose is fulfilled.</li><li>One of the success criteria could be that the purpose should be fulfilled within a specified deadline.</li></ul>
B	**What, who and how are we now?**	In addition, a thorough mapping of the current situation for the FM organisation should be undertaken, including which competences the organisation possesses and which collaborative relationships the organisation has internally in the company and with external providers and collaboration partners.
C	**How should we be organised in the future?**	After mapping the current situation, proposals for potential future organisational plans are formulated. The proposals describe changes in, for instance, work tasks, staff and collaborative relationships. This typically results in two to three alternatives, which you analyse in detail and compare advantages and disadvantages. This is the basis for a decision on a new organisational plan.
D	**Need for knowledge and competences**	Based on the new organisational plan, the need for new knowledge and competences is analysed. This involves clarifying the following: <ul><li>Eventual redundancies, replacements and continuing education among current staff</li><li>Need to employ new staff</li><li>Changes of providers and external consultants</li></ul>
E	Implement	The new organisational plan is implemented as a change process with the involvement of all staff. The head of the FM organisation must lead and explain the reasons for change in the organisation and point out the advantages it will provide, as well as the challenges that you expect to face in the implementation process. The managers on all levels should undertake a process together with their staffs to clarify uncertainties and decide how to deal with problems that appear during the implementation.

Phase	Name	Description
F	**Follow-up and adjust**	During and after the implementation, the management of the FM organisation should follow up on the organisational change and decide on necessary adjustments of the implementation plan and the organisation in such a way that the purpose for conducting the process is fulfilled.

Tools for strategy development

Figure AII.1 shows the names of four tools that can be used for organisational design in FM and the phases in the process where they are most relevant. A short description of each tool is given next, and this is followed by a section about each tool.

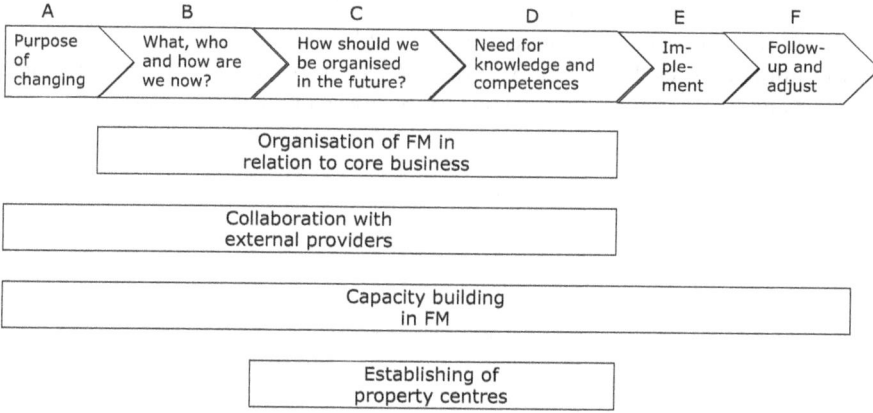

Figure AII.1 Tools for organisational design and the process phases

Organisation of FM in relation to core business

The tool helps you to analyse the decision competences and centralisation/de-centralisation in your current forms of collaboration in phase B, and it can further be used to support decisions about how you should be organised in the future and the need for new knowledge and competences in phases C and D.

Collaboration with external providers

The tool provides you with an overview of advantages and disadvantages with different ways to collaborate in a standardised and a non-standardised production, and with singular as well as ongoing tasks with and without time limitations. The systematic can be used when you analyse your current and future organisational design.

Capacity building in FM organisations

This tool is for you if you want to be the champion to lead your organisation upwards on the FM maturity stairway (janitor, building owner, active building owner, preferred collaboration partner, innovator). The model provides insights into what it takes to build capacity, independent of how high you intend to raise your organisation. To lead the development, you need to be able to motivate and support your staff so that they at any time know what the right thing to do is. The tool focusses on the staff and provides you with ideas for principles and approaches, which can pave the way towards more self-management.

Establishing of property centres

Several organisations decide to establish a central FM function, and the management task consists of establishing a new organisation with a shared picture of the mission, task portfolio, service levels, customer relations, economy, organisation, etc. This tool is for you if you are going to establish a property centre. The tool was developed based on experiences with establishing centres in a number of Danish municipalities. It provides you with advice in terms of seven steps to the development journey that characterises establishing a new property centre.

A.5 Organisation of FM in relation to core business

What?

The tool is a model for the organisational relationships between core business and the FM function. The model comprises different principle models for organisational coordination. It provides a terminology and an analytical framework you can use to clarify the organisational relationships between support functions and the core business, and evaluate which forms are most appropriate.

Table A.3 shows which forms of coordination between different units in an organisation that typically are the most appropriately dependent on whether decision-making is one-sided, two-sided or multi-sided and whether it takes place centrally, on an intermediary level or de-centrally.

Table A.3 Forms of coordination related to centralisation and decision-making

Degree of centralisation decision-making	Centralised	Semi-centralised	De-centralised
One-sided	Authority relationship	Agent relationship	Norms/customs
Two-sided	Partnership	Negotiation	Price
Multi-sided	Coalition	Voting	Team

Why?

You can use the model to clarify the collaborative relationships and forms of coordination between external providers, internal FM and core business, and as help to decide sourcing questions. The model provides a language and some categories that can support the decision-making process. When using the model, it becomes clear how FM can be organised in relation to support and dialogue with the core business. It helps to shed light on different ways to organise FM to optimise the use of resources in the tasks involved in supporting the core business. There can be differences in the degree of outsourcing and different weighting in how close the support functions are to the core business.

Who?

The tool is mostly aimed at decision makers who shall ensure that the organisation is shaped and dimensioned appropriately – i.e. heads of FM and FM middle managers – and it is used as a basis for their dialogue with top corporate management.

How?

The persons responsible for developing and dimensioning the FM organisation use the tool as a frame of understanding and as a basis for analysing the existing relationships between the internal FM organisation and the core business on one side and between the internal FM organisation and external providers on the other side. The focus is on the relationships, both on the strategic and operational levels. It can also be advantageous to include the relationships on the tactical level (see Chapter B.27, "Value-Adding Management").

At the strategic level, it will often be appropriate with a multi-sided, central coordination in terms of a coalition, where the management of FM engages in dialogue and shared decision-making with all the main stakeholders in corporate management. This is because the focus is on accommodating the long-term needs for physical frames and infrastructure for the whole corporation. Thus there is a need for FM to have a business orientation at the strategic level.

At the tactical level, it will often be appropriate with a two-sided form of coordination in between the central and de-central level. This can take the form of bilateral negotiations with representatives from the different business units or departments of the core business. The aim is to accommodate each their specific needs for space and services in a medium-term perspective. Thus there is a need for FM to have a customer orientation on the tactical level.

At the operational level, it will also often be appropriate with a two-sided form of coordination. This is most likely to be de-central between FM and the end users based on price as coordination mechanism, for instance price per order or as a part of internal rent or service charge. Thus there is a need for a service orientation at the operational.

Related tools

The tool can be used together with *collaboration with external providers* (see Chapter A.6) and *adaptation between FM product and process* (see Chapter A.2).

Further information

The tool is presented in more detail in Chapter B.31.

A.6 Collaboration with external providers

What?

The tool is based on an investigation of how different forms of collaboration with external providers influence the provision of FM services concerning productivity, quality, earnings, innovation, knowledge sharing, trust and conflict solving. The investigation made a fundamental distinction between FM provisions characterised by standardised production and non-standardised production.

Table A.4 shows the principle forms of collaboration that are most suitable for the standardised and non-standardised provision of FM, respectively, and the forms of knowledge sharing and management technology that are most appropriate for the two situations.

Table A.4 Characteristics of standardised and non-standardised production

Performance	Standardised production	Non-standardised production
Form of collaboration	Contractually regulated	Relational capabilities • Among companies and among persons • Flexible forms of collaboration
Knowledge sharing	Formal education Explicit knowledge	Competences and tacit knowledge • To be able to solve the task • Responsibility • Trust • To understand customers' and users' needs • Service mind-set • Knowledge sharing and experience building
Technology	SLA, KPI, administration and control	BIM and visualisation, embedded technology

Why?

The tool provides a research-based foundation to make decisions about which forms of collaboration FM organisations should arrange with external actors, including whether it is appropriate to enter into long-term collaborations, or if it is more appropriate to base the collaboration on singular tasks or shorter, time-limited collaborations. By using the tool, you will obtain a deeper understanding about what is at stake in different types of contracts and forms of collaboration.

Who?

The tool is aimed at the main decision makers who shall ensure that the organisation and provision of FM function is appropriate. Therefore, it is of value for most facilities managers with management responsibility and staff on the strategic level involved in sourcing analyses.

How?

The tool distinguishes between the following four forms of collaboration with external providers:

- One-time tasks
- Continued tasks, without agreed upon time limits
- Continued tasks, with agreed upon time limits
- Operational partnerships

Operational partnerships are the most advanced of these with regard to long-term collaboration based on close relationships between the client and provider. An agreement with shared goals for the collaboration and service delivery in relation to process and optimisation of competences has typically been made. Dialogue, openness and trust are in focus. The shared goals for the service provision are based on a demand specification with follow-up on mutual clarification and specification of the expectations for the outcome of the service delivery. The goals can include user satisfaction, professional targets, development of methods, economic targets, service targets, etc. A model for solving conflicts is typically included based on conflicts to be solved by dialogue at the level where the conflicts occur and with a possible escalation route to higher management levels.

For singular tasks, none of these elements are normally included, but they can be included in various degrees in continuous tasks.

In the analysis of sourcing and potential external forms of collaboration, you should evaluate the following:

- The character of the FM service that the company needs
- To which degree the elements mentioned earlier should be included in the collaboration with external providers

- To which degree knowledge sharing and what use of technology for management of the collaboration should be applied

Related tools

The tool can be used together with *organisation of FM in relation to core business* (see Chapter A.5) and *adaptation between FM product and process* (see Chapter A.2).

Further information

The tool is presented in more detail in Chapter B.19.

A.7 Capacity building in FM organisations

What?

The model provides insight into what it takes to build capacity in a FM organisation with the aim of individual members of staff knowing what the right thing to do is at any time based on some overall criteria for well-conducted work.

Figure A7.1 shows the model with the elements that are important to discuss to be sharp on what it takes to build strategic capacity in the FM organisation. A clear strategic goal is a precondition for long-term planning and makes it possible to formulate clear criteria for well-conducted work (quantitative KPI's/qualitative values).

The individual facilities manager has, besides clear goals for the future organisational development, a need for explicit knowledge of different topics. Furthermore, a number of organisational- and planning-related preconditions have to be in place to make it possible for staff members to actually experience control of their own practice.

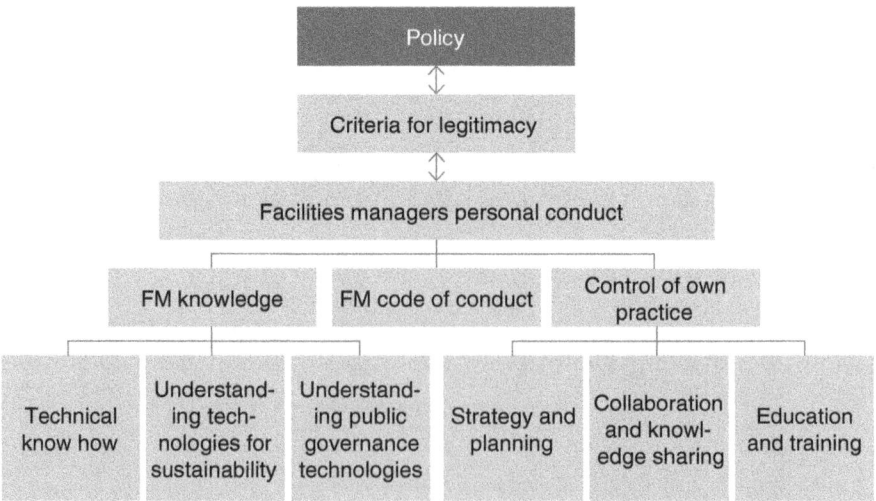

Figure A7.1 Model for capacity building in FM organisation

Finally, a shared code of conduct for the FM organisation can support a culture where the focus is on, for instance, service, quality, coordination, collaboration – or whatever is seen as important to reach the strategic goals of the organisation.

Why?

The model can be used as a starting point for a dialogue concerning capacity building in the FM organisation. This dialogue should focus on creating the frames that are needed for the individual member of staff to live up to the increased requirements for self-management to a higher degree. Thereby, the staff will be able to contribute actively to the strategy development. The goal is that the organisation is geared to work strategically towards a higher degree and thus become an even collaboration partner for the other parts of the corporation and to external collaboration partners.

Who?

The model is relevant to use as a starting point for dialogue in immature as well as in fully professionalised FM organisations. In organisations with a lack of coordination of the activities, experience of unclear goals, uncertainties in relation to roles, etc., the model can contribute to a focus on where there is a need for more effort to increase the total organisational capacity. In mature, well-managed FM organisations, the model can be a starting point for a dialogue about where the organisation should be adjusted to implement long-term strategic goals. This can, for instance, be goals about contributing to a circular economy, which requires new knowledge and competences, and a new code of conduct in relation to procurement and waste handling.

The model can be used by the management team in the FM organisation and as a central element in the dialogue among the staff facilitated by development experts/consultants who have an assignment to build capacity in the organisation to be able to work more strategically. It can be relevant to involve an external consultant in case the challenges in the operation are caused by problems that are hard to verbalise if they are internally employed.

How?

The model can be used as a common frame of reference and as a starting point for dialogue with the aim of focussing on whether there is coherence between goals and organisational capacity. The dialogue can start with questions like the following:

- Do we have a clear strategic aim with our work? Do the management and staff in the FM organisation experience the goals as meaningful?
- Is the strategic aim translated into clear criteria for well-conducted work? Do the criteria support a holistic view and provide an understanding of the

purpose with the tasks? If yes/no, what does that imply for the individuals' task solutions?

- Do we have the right knowledge and competence in the organisation? Now and in relation to future strategic goals?
- Do we have a sufficient planning horizon to make credible budgets and ensure a coordination of activities across disciplines?
- Do we have a sufficient focus on knowledge sharing and codification of explicit knowledge? Are we vulnerable to the loss of knowledge if key staff disappears, and what can we do to attain tacit knowledge?
- What do our customers and collaboration partner experience when they collaborate with us? Would it be of value to describe it in a common code of conduct?

The dialogue can eventually be supported by the facilitator graphically documenting the input from the group in a large print of the model with a picture of an archetype of a staff member placed in the middle of the picture. In this way, it becomes clear that the code of conduct of a facilities manager depends on a lot of factors, and some of these lie outside the individual's immediate sphere of influence.

Related tools

The tool can be used together with *energy efficient* FM (see Chapter A.21).

Further information

The tool is presented in more detail in Chapter B.7.

A.8 Establishing of property centres

What?

The tool consists of seven steps that can help in establishing property centres. This recommendation is based on municipalities in Denmark, which have been through a process of going from de-central to central property management. This means they have obtained experiences from forming a united organisation with competences to give advice and operate buildings in the whole life-cycle phase from planning, new building, administration, optimisation, operation and dismissal. The seven steps are formulated in overall terms and point to the management challenges in creating a new and united organisation.

Table A.5 shows the seven steps to help in establishing property centres, which have been formulated based on the experiences and recommendations from Danish municipalities studied in a research project.

Why?

It is not a simple task to create or reorganise a FM organisation because any way of organising has its advantages and disadvantages, and even though there will be people advocating for a central FM unit, there is also likely to be resistance to

Table A.5 Seven steps to establish property centres

Step	Description
1	Start with the FM provisions that you can agree on
2	Formulate a strategy for staff information and involvement
3	Make a plan early for the future servicing of schools/institutions, which earlier had their own FM organisation/staff
4	Engage external expertise if you lack time or competences
5	Create an easy access for users
6	Define an appropriate service level for all buildings
7	Establishment of a property centre is an ongoing development process and probably never ends

change. As one of the new managers of a property centre, you need to motivate your staff, and you will be in a more favourable position if you know the challenges that others have experienced when they were in a similar situation. You need to be a strong negotiator so that from the outset you make an appropriate agreement concerning the new FM organisation's tasks, service levels, organisational structure and economy.

Who?

If you are involved in establishing a new, central property centre, then this tool is for you and your management colleagues when you discuss goals, frames and procedure in a helicopter perspective.

How?

The procedure is also known as 'back-casting':

- Describe where you want to be in five years. What characterises the property centre and its activities? What successes show that you have reached your goals?
- Where are you today? What characterises your current organisation and its activities? What advantages and disadvantages do the current way of organising have?
- Identify the changes that are needed within the next five years and the initiatives that can lead you to get there. Consider the order and make an implementation plan.

When you have outlined your own process towards a central property centre, you should make a quality check of your plan by using the seven steps. Have you to a sufficient degree considered the challenges that the municipalities have experienced? Read more about these challenges in Chapter B.33.

There may be a need to elaborate on your plan and specify in more detail how you will implement a strategic decision on establishing a central property centre and at the same time succeed in getting a good start on the operation of the new centre.

Related tools

The tool can be used together with *capacity building in FM organisations* (see Chapter A.7).

Further information

The tool is presented in more detail in Chapter B.33.

A.III

Space planning

Opening dialogue

Evelyn, project manager and developer, to Karen, building project manager:

- ¡Division West is going to employ more staff, and the management has changed the organisational structure, so we have developed a proposal for a new layout of their building. It requires a major churn of staff and perhaps rebuilding.

Karen's response:

- I will start making a building project proposal.

Why this process?

The purpose of the process is to ensure that the capacity and allocation of space is in accordance with the company's current and expected future need of space of different categories. This also includes an appropriate location, distribution and utilisation of space.

What triggers starting this process?

- When a major reorganisation in the company occurs, including a merger with another company
- When a part of the company needs to extend or retract activities with a related change in the space needed
- When it has been decided that the company must reduce property costs or introduce new principles for workplace layout
- When there is a need for major renovation of a building

The process

The process for space planning consists of eight phases, which are described in Table A.6. The eight phases can be related to the four phases of Plan, Do, Check and Act in the commonly used Deming quality circle, as shown in Figure AIII.1.

Table A.6 Phases in the process of space planning

Phase	Name	Description
A	**Evaluate supply and demand for space**	As a starting point, you need to create an overview and estimation of the current and expected future need for space. This is often based on information collected from the different user departments about expected changes in their needs for space. By comparing the records of the company's current space and the expected future needs, you map the mismatch between supply and demand divided in relevant space categories and calculate the need for change.
B	**Identify alternatives**	Based on the estimation of the need for change, you make a further analysis of needs (space briefing), and from this, you identify possible solutions to accommodate the needs. It can, for instance, include redistribution, change in degree of utilisation, buying/selling, renting/letting, establishing temporary buildings or building/rebuilding projects.
C	**Evaluate alternatives**	The most favourable alternatives are analysed further by the FM organisation and compared in a decision proposal. The criteria for comparison can be based on the corporation's property/space strategy if such a strategy has been formulated.
D	**Decide**	The managers of the affected departments are informed about the decision proposal, and based on their input, corporate management decides which of the alternatives should be implemented.
E	**Plan**	When a decision has been made, the FM organisation starts the detailed planning of the solution in collaboration with the affected departments. If the decision is a building project, then the process is changed accordingly, and it will proceed as described in A.IV.
F	**Implement**	The plan is implemented by the FM organisation in close collaboration with the affected departments and external partners, such as building companies, removal companies and providers of furniture and equipment.
G	**Evaluate results**	After implementation is finished, the result is evaluated. This can, in the simplest form, be a dialogue between the FM organisation and managers from the affected departments about their satisfaction with the process and the new space solution, but in case of more comprehensive projects, it can be relevant with a questionnaire survey among the affected staff.
H	**Reconsider**	The FM organisation updates the records on space and needs, and reconsiders the match between supply and demand and the space utilisation.

Tools for strategy development

Figure AIII.1 shows the names of four tools that can be used for space planning in FM and the phases in the process where they are most relevant. A short description of each tool is given next, and this is followed by a section about each tool.

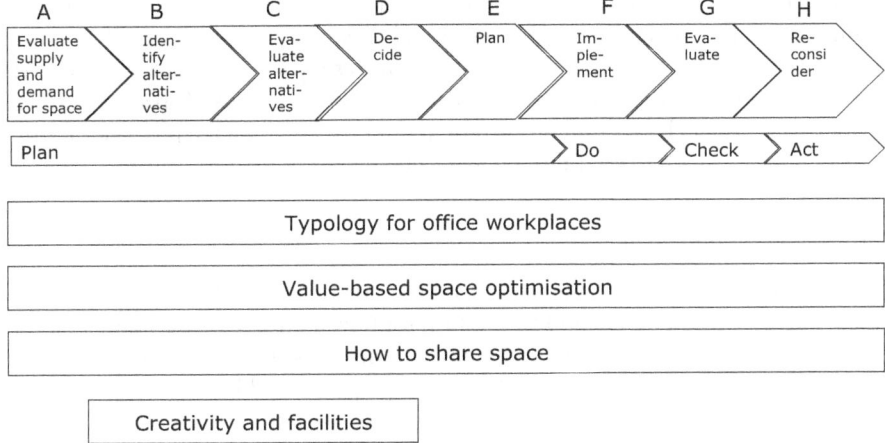

Figure AIII.1 Tools for space planning and the process phases

Typology of office workplaces

If your facilities are used as office workplaces for knowledge workers, then this tool can be useful. It includes a typology of both traditional and more experimental office workplaces, and it provides you with a common terminology and a detailed and systematic view on the types of workplaces that exist around the world and at your organisation. The typology can be used both in the early analyses during decisions about new office layouts and during planning, implementation and evaluation of changes in workplaces.

Value-based spaces optimisation

Optimisation of space should take its starting point in the values and strategies of the organisation. Space optimisation can firstly include qualitative optimisation so that the space is designed to better support the activities of the organisation – for instance, concerning more creativity as described later in the tool creativity and facilities. Secondly, it can include quantitative optimisation so that space is utilised better, for instance, by sharing space between different groups of users, as described in the next tool: how to share space. The tool on value-based space optimisation can help you work with space optimisation more generally. It builds and combines different former methods and is divided into seven steps. Together, the seven steps cover all phases in the process of space planning. The tool was developed in relation to gymnasiums (higher-education institutions) in Denmark, but it can also be used in other types of organisations.

How to share space

If the users of your space are going to share facilities – workplaces or other functions – then you can use this tool as a benefit. It provides a detailed view of different ways that you can share space and makes it easier to communicate to the rest of the organisation. It can be used in the whole process of space planning – from phase A to H – to analyse current and future needs, make proposals for new space solutions and plan and implement changes.

Creativity and facilities

Finally, you can benefit from using the tool creativity and facilities, which provides an analytical model and inspiration for how a creative environment can be supported. It can be used when you develop new possible solutions and evaluate and decide what you want to implement in phases C and D.

A.9 Typology of office workplaces

What?

The typology is a status for how we organise and design knowledge workplaces around the world and in Europe. The research includes a re-evaluation of 30–40 years of experimentation, research and debate on workplace concepts, such as remote working, open offices and flexible workplaces. It describes the experiences with 'alternative offices' and 'new ways of working' (NWOW), and points to trends in the design of workplace concepts that challenge the traditional perception of the office.

The ten types of office workplaces are shown with illustrative drawings in Figure A9.1. An extended version with names and descriptions is included in Chapter B.1.

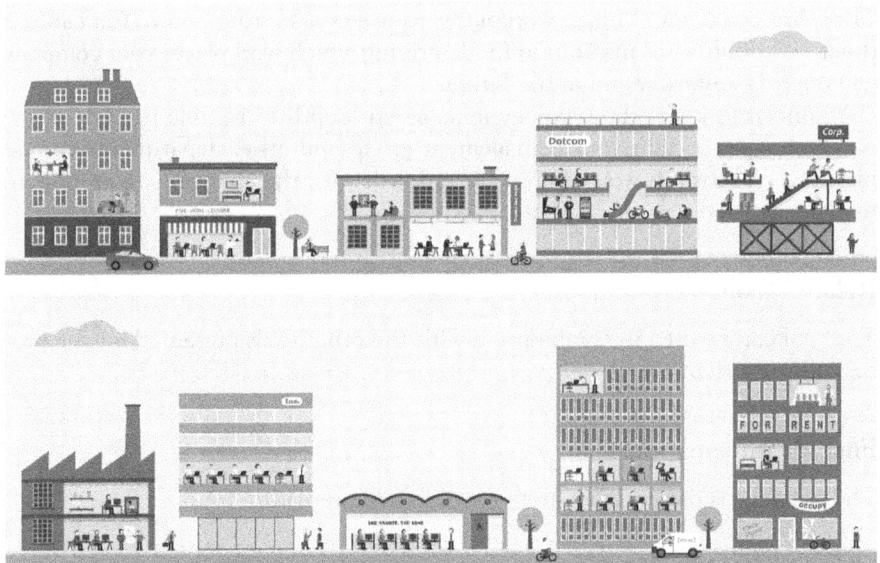

Figure A9.1 Typology of office workplaces

Why?

You can use the tool for – at least – two purposes.

1   The typology provides your FM organisation with a common terminology and a structure in your internal work on analysing your current supply and demand for workplaces. You obtain a basis for conducting such an analysis and making decisions – and implementation – grounded in current knowledge about knowledge workplaces.
2   You can also use the tool when you need to enter into a dialogue with the departments that are going to be subject to a change in workplaces about the decision, planning, implementation and evaluation of the office reconfiguration. Where and how do you want to create the frames for different types of knowledge work?

Here we will focus on description of the use in relation to evaluation of supply and demand.

Who?

The tool is of particular relevance for you if you have the task of investigating which spaces your company has at its disposition and preparing a proposal for which spaces you will need in the future.

How?

There are ten types of office workplaces to use as a starting point. You can use them as they are or as inspiration for describing which workplaces your company will offer its staff, now and in the future.

You need to know the typology in sufficient depth to be able to present the different types so that your management group and other departments can be involved. Afterwards, you can specify a project with the changes you propose to implement.

Related tools

The tool can be used in combination with the other tools presented for the process of space planning.

Further information

The typology is presented in more detail in Chapter B.1.

A.10 Value-based space optimisation

What?

The tool is a procedure for a space optimisation process where you can involve users in an evaluation of the existing physical frames to identify goals and make plans to improve the facilities. This is important to ensure that the users are sufficiently satisfied with the space. The tool is a space manager's guide to specify renewal projects through a dialogue with stakeholders.

Table A.7 shows the seven steps, which are recommended for use in value-based space optimisation.

Table A.7 Seven steps in value-based space optimisation

Step	Objectives	Activities
1	Clarification of purpose and success criteria	Interview the head of school and the facilities manager
	Identifying stakeholders	Walk-through with the head of school and the facilities manager
	Preparing project plan and clarify resources	
	Collection of data about the organisation, the buildings and the space challenges	
2	Collection of data about the use of space	Observations and interviews
		Analyse space utilisation
3	Discussions about the existing use of space	Focus groups with the primary stakeholders
4	Clarifying which space solutions work well and which do not – generally and related to specific aspects of the analysis	Walk-through with the primary stakeholders
5	Involvement of a larger group of stakeholders	Questionnaire survey
6	Preparation of proposals for space optimisation and implementation plan	Workshop with the primary stakeholders, the head of school and the facilities manager
7	Implementation of space optimisations	Churns, rebuilding, etc.

Why?

Over the years, most buildings have a need for updating because the work processes have changed or there is a need for adaptation to fewer or more users. The renewal can take place area by area or as a major, comprehensive project. This tool is for those who face a renewal of layout or a minor rebuilding to optimise space and want a process in which stakeholders can be involved. The process should ensure the satisfaction of the end users in relation to a specific space and align supply and demand. By using this tool, you can collect input about the need for updating, which only the users can provide you.

Who?

An external consultant can facilitate the process, but the facilities manager must ensure that the process takes place and that sufficient time and resources are used during the dialogue. The future users should as much as possible be involved in the walk-throughs and workshops.

How?

The process starts by defining goals and success criteria for the coming renewal and preparing the following data collection. The second step is the data collection concerning the current use of the space. Based on that, you identify particular well-functioning areas and particular problem areas with a small group of users in a walk-through, followed by a workshop and broader data collection with a questionnaire survey. You prepare a proposal for space optimisation, which makes the layout more in accordance with the organisation's values and activities.

Related tools

The tool can be used in combination with the other tools presented for the process of space planning as well as the tool *usability briefing* (see Chapter A.13).

Further information

The tool is presented in more detail in Chapter B.4.

A.11 How to share space

What?

The tool consists of a typology that gives an overview of the different degrees of shared space and how to characterise a shared space. The characteristics concern what to share, why to share, when to share, who to share with and how to share. In addition, a *Guide to Shared Space in Municipalities* has been published separately both in English and Danish as a supplement to the typology.

The tool shows how to work with the sharing of space in practice. It was developed based on examples from both private and public organisations.

Figure A11.1 shows the typology for space sharing starting with no sharing and with three different degrees of sharing. An extended version is included in Chapter B.5.

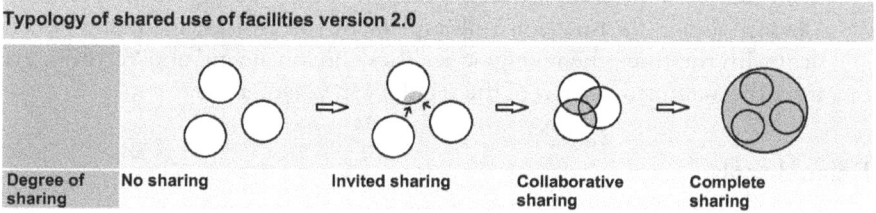

Figure A11.1 Typology of the shared use of space

Why?

The tool supports your work with achieving a more optimised utilisation of the facilities and space your organisation has at its disposal. This can have consequences for economy, environmental impact and user experience. The typology provides you with a common terminology and a systematic approach to the internal work of analysing the current supply and need for space. The *Guide to Shared Space in Municipalities* can afterwards be used as a guideline to implement space sharing – for instance, across departments, companies and user groups.

Who?

The tool is of particular use for you if you have the task of investigating the space your company has allocated and you are making an estimation of the space you will need in the future. This typology focusses on how you use your space and whether you can create new frameworks for collaboration with internal and external partners.

How?

You can use the typology to:

- Identify intentions to share
- Draft a project based on the intentions
- Realise the project in practice with help from the typology

When you are going to create more shared space, it is not only a specific practical and physical process but also it will often be a cultural change project. Chapter B.5 describes the typical advantages you can expect, as well as the typical challenges you need to address.

The process of sharing typically take place in the following three steps:

1 Clarify how much and how you want to share – i.e. 'invited sharing', 'collaborative sharing' or 'complete sharing'.
2 Describe your shared space by using characteristics: 'what', 'when', 'why', 'who' and 'how' from the extended version of the typology in Chapter B.5.
3 Develop a plan for how you will implement the project, including how to deal with the three themes: 'practicalities', 'involvement' and 'territoriality' from the extended version of the typology in Chapter B.5.

Related tools

The tool can be used in combination with the other tools presented for the process of space planning.

Further information

The tool is presented in more detail in Chapter B.5.

A.12 Creativity and facilities

What?

The framework is based on six case studies which describe different ways to establish and support creative environments, both within private and public companies and in city areas. There are four different aspects in facilitating creative environments: 'facilities', 'facilitation', 'use' and 'culture'. Thus the development of innovative organisations should not only be seen as a question of competences but also as a broader development of experimental and learning practices, which again requires management prioritisation and support. Creative facilities are not sufficient. Facilitation is not sufficient on its own. A change in culture is needed.

Table A.8 shows a number of central factors listed as dichotomies, which can be used in an analysis of the basis for developing and facilitating creative environments.

Table A.8 Dichotomies in relation creativity and facilities

Either?	Or?
Silo thinking	Cross-disciplinary
Zero-mistake culture	Experimental approach
'Participation'	'Consumption' of planned offerings
Formality, planning	Openness, temporality
Top-down management	Bottom-up management
Personalisation	Corporate politics and brand
Effectivity	Reflectivity

Why?

The discussion regarding creativity during the last decades has had a huge influence on the development of cities, private and public companies and society at large. Many see creativity as a key to a better future, and they emphasise that the creative economy has the potential to both generate growth and strengthen social inclusion, cultural diversity and human development.

The tool provides a systematic which you can use as a starting point for reflections and discussions of the development of creative environments in both city areas and companies.

Who?

The tool can be used by space managers, staff involved in development and consultants who have been given the task of developing and supporting creative environments either in a city or a company context.

How?

You make an analysis of the organisation's fundamental attitude towards creativity and creative environments so that you get to know which buttons to turn if you want to go in new directions. You can use the six cases described in Chapter B.2 and their variations concerning facilities, facilitation and use and culture in your considerations and formulations of possible goals for the initiatives to increase creativity in the organisation. The six cases reveal a number of contradictions. With your company's mission and vision as a starting point, you can use the dichotomies in Table A.8 to verbalise the shared imaginations and desires of the stakeholders.

Related tools

The tool can be used in combination with the other tools presented for the process of space planning.

Further information

The tool is presented in more detail in Chapter B.2.

A.IV

Building project

Opening dialogue

Karen, building project manager, to Carl, FM team leader:

- We are going to plan the new building towards the east. Who can we involve from your team to ensure that considerations for building operation are integrated in the project?

Carl's response:

- Our commissioning coordinator will contact you, and you have all our written requirement standards.

Why this process?

When you are planning new buildings, it is important that you make use of the knowledge and experiences from the FM organisation and from the users. It is about making decisions that are as durable over the long-term as possible so that the finished buildings fulfil the short and long-term needs of the company to the highest possible degree with regard to usability, operational friendliness, life-cycle economy and branding. This process takes all of these things into account.

What triggers starting this process?

- When the decision to start a building project – independent of it being a new building, rebuilding, renovation or refurbishment – is derived from the process of space planning

The process

The process for a building project consists of eight phases, which are described in Table A.9. The eight phases can be divided into pre-project, project and post-project, as shown in Figure AIV.1.

Table A.9 Phases in the process for a building project

Phase	Name	Description
A	**Initiate building project**	Prior to starting a building project, a feasibility study is often carried out to clarify possible alternative solutions, resulting in a decision proposal provided to corporate management.
B	**Strategic briefing**	If an initial decision to go on with a building project is made, a strategic briefing is usually done in the pre-project stage to identify the overall vision, goals and intentions with the project, including involvement of top management, key persons among the staff and managers and specialists from the FM organisation.
C	**Building briefing**	The actual building project starts with the building briefing, where the specific requirements of the building are defined. This often involves a number of work groups with members of staff and internal and external technical experts, including representatives from the FM organisation. The building briefing is managed, facilitated and coordinated by the company's building client function, which can be a part of the FM organisation – eventually together with an external client consultant.
D	**Design**	During the design phase, the building's design and technical features are developed gradually in more detail during a number of sub-phases, and the early design may start in parallel with the detailed building briefing. The design typically is developed by external architects and consulting engineers who have an ongoing dialogue with representatives from the client, FM organisation and users.
E	**Construct**	During construction, the physical realisation of the building takes place. This is typically done by contractor companies that are in an ongoing dialogue with the client, site management and supervisors from the design team.
F	**Handover**	During handover, the building is formally transferred from the contractors to the client, together with a record of faults which the contractor has to amend. At the same time, the responsibility for operating the building is allocated to the operational part of the FM organisation.
G	**Operate**	In the start of the post-project stage, the FM organisation takes all the necessary utility systems and other technical systems into use and installs the user organisation's equipment and furniture so that the company can occupy the building.
H	**Evaluate**	It is recommended to arrange an evaluation of the users' and the operational staff's satisfaction with the building within one year after occupation. Unfortunately, this is often not done, but it is a good idea so that necessary amendments can be made and experiences collected for upcoming building projects.

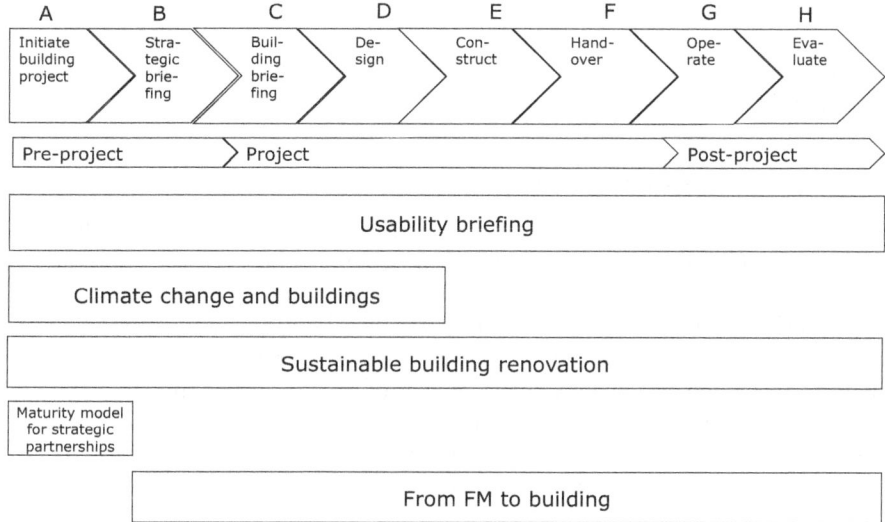

Figure AIV.1 Tools for building project and the process phases

Tools for strategy development

Figure AIV.1 shows the names of five tools that can be used for building projects and the phases in the process where they are most relevant. A short description of each tool is given next, and this is followed by a section about each tool.

Usability briefing

When you are going to involve users, this tool can become useful. It describes how you involve users in an appropriate way: how they should be involved and what they should focus on. The tool covers the whole building project, but the main emphasis is on the early phases.

Climate change and buildings

This tool provides a method for quantitatively estimating costs/benefits of implementing measures in building projects that at the same time aim for improving the building's resilience in relation to climate change and sustainability of the environment and image. The tool can, in particular, be used in the pre-project phases and the early project phases – phases A–D.

Sustainable building renovation

During planning and management of building renovation with a focus on sustainability, the tool, also called RENO-EVALUE, can be helpful. It can be used as a basis for dialogue and management of expectations in the early phases. You can also use it to evaluate the project during and after the execution regarding the satisfaction of different stakeholders with the project and their views on the fulfilment of the project goals. Thus the tool can be used during pre-project, project and post-project.

Maturity model for strategic partnerships

In the building industry, building and renovation projects are increasingly procured by the use of frame agreements covering a portfolio of projects instead of procuring one project at a time. The aim is to create better and cheaper buildings by establishing a strategic partnership between a building client and a consortium consisting of companies from the supply chain. However, there are high requirements for the companies that want to be part of such a partnership. To evaluate the maturity of one's own company and potential collaboration partners, you can use this tool. The tool can primarily be used in the initial phase A prior to or in connection with initiating building projects.

From FM to building

New buildings are often not as user and operational friendly as one would expect. This is a well-known problem. Involvement of knowledge and experience from FM in building projects is an obvious possibility for preventing this problem. This tool can be used to evaluate which mechanisms are most suitable to transfer knowledge from FM to the building project. It consists of a typology of mechanisms to transfer FM knowledge – both as requirements during design and as performance during construction. The tool can be used from strategic briefing, phase B in pre-project, during all project phases C–F and until evaluation, phase H in post-project.

A.13 Usability briefing

What?

The model shows how you can involve users and utilise evaluations in connection to briefing and design during the whole building project. You can use the model as a basis for dialogue with the organisation when you are planning user involvement in building projects.

Figure A13.1 shows the model for usability briefing divided into the typical phases of a building project and the four main activities. An extended version is included in Chapter B.2.

Why?

The purpose is to ensure that the building fulfil users' needs to a high degree. Many projects only involve users in the early phases, but this model is based on a continuous involvement of users, systematic use of evaluations and continuous briefing.

Who?

The model can be used by project managers of building projects as a planning and management tool to help in organising the building project and is typically used as a basis for dialogue with other stakeholders in the beginning of the planning.

How?

You use the model in the early phases of a building project to plan user involvement. It provides input into the activities that should take place, who should be involved and what specific tools should be applied. The model provides you with both an overview over the process and an indication of when the user involvement should be most intensive, but it also gives you concrete suggestions for tools that can be applied (meetings, brainstorms, workshops, mock-ups, etc.) and who should be involved at which stage.

Related tools

The tool can be used in combination with *from FM to building* (see Chapter A.17) and *value-based space optimisation* (see Chapter A.10).

Further information

The tool is presented in more detail in Chapter B.3.

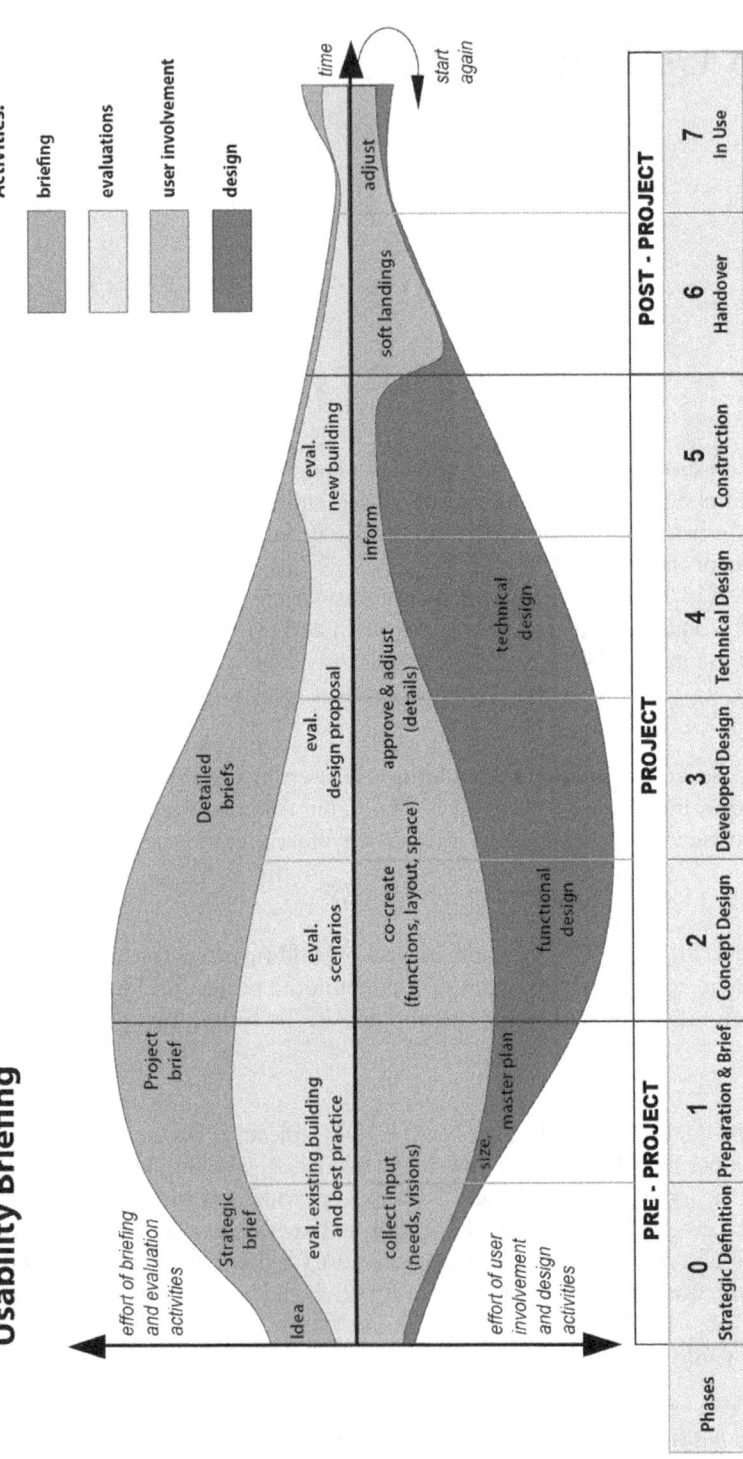

Figure A13.1 Model for usability briefing

A.14 Climate change and buildings

What?

This is a method for estimating the total cost of measures to prevent climatic impacts – for instance, measures to collect, percolate or delay rainwater. The method is an analysis and decision support tool which combines and quantifies climate resilience and sustainability (environment and image).

Figure A14.1 shows relationships between climate impacts, climate change and buildings. The method provides a basis for decisions about choosing relevant measures to prevent climate impacts and their consequences for buildings.

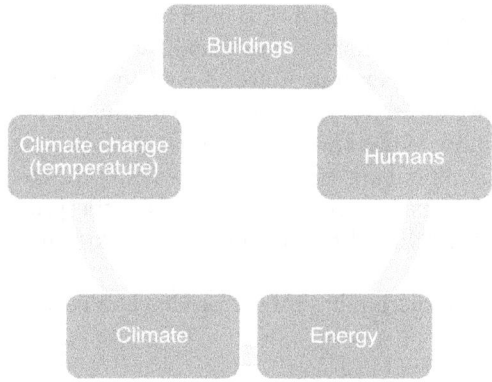

Figure A14.1 The relationships between buildings and climate change

Why?

When you plan to invest in environmental- and climate-related measures, it is often expected that you prepare a business case, which compares advantages and disadvantages, and further provides an evaluation of the probability for the best/worst scenario to occur, and what the consequences will be. This method provides a basis for the evaluation of the measures which you implement to secure your building against future climatic impacts, such as flooding and increasing temperatures.

The special thing about the method is that it both takes into account whether a measure is appropriate in relation to the challenge it is supposed to prevent and looks at if the measure in itself contributes to an increase in the climate impact.

Who?

The method is aimed at you if you have been given the task of preparing a basis for decision and proposals for environmental- and climate-related measures. It can also be used in a broader way as an analysis and decision support tool by managers and staff responsible for sustainability and related consultants.

How?

The method consists of seven steps, which include evaluation and calculation of risks in relation to climate impact, sustainability and resilience. In the last step, you choose a solution based on a cost/benefit analysis. Before starting to work with the method, you should make a plan for the analysis and consider who should contribute knowledge input to the different steps. The seven steps are as follows:

1   Determine the resilience of the building to the disturbance – e.g. at what temperature the building's function will be compromised.
2   Determine the costs associated with both the loss of the building's functionality and the building's current sustainability.
3   Determine the corresponding probabilities associated with each cost.
4   Determine the expected costs associated with the current resilience and current sustainability of the building using risk analysis.
5   Determine the capital and operational costs of each remedial solution.
6   Determine the expected cost associated with the proposed resilience and proposed sustainability of the building using risk analysis for each remedial solution.
7   Select (or not) a solution based on cost-benefit analysis.

Related tools

The tool can be used in combination with *strategies for sustainable* FM (see Chapter A.3).

Further information

The tool is presented in more detail in Chapter B.8, including an example for using the method.

A.15 Sustainable building renovation

What?

The tool is also called RENO-EVALUE and can be used to formulate holistic goals and evaluate the sustainability of building renovation projects. It is designed with four categories, out of which three represent social, environmental and economic sustainability, while the last category concerns project organisation. Each category is divided into two parameters: social sustainability in process and product, environmental sustainability in climate and resources, economic sustainability in value and money (Euro/Kroner) and project organisation in consultant/contractors and developer/client.

Figure A15.1 shows an example of the results of a sustainability evaluation of a building renovation based on RENO-EVALUE illustrated as a radar diagram divided into the eight parameters and a scale from 1 (lowest) to 5 (highest). For the sake of simplicity, the results of only one person's evaluation is shown, but the intention is that evaluations should be made of all the important stakeholders and compared.

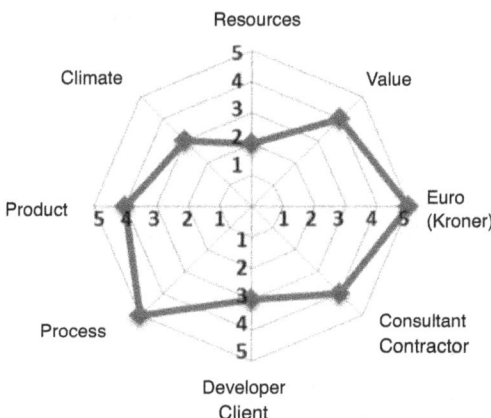

Figure A15.1 Example of results from RENO-EVALUE

Why?

You can use the tool to support a dialogue-based decision process in the early phases of renovation projects. During the project, the tool can be used in an ongoing evaluation of the achieved results of the project – compared with the defined goals and identified expectations. The tool not only focusses on the result of renovation – the product, but also covers the process of renovating.

Who?

Decision makers among building clients, housing associations, property administrators and facilities managers, etc., can benefit from using the tool, even if they do not possess the necessary technical competences to evaluate renovation projects in detail. The tool can be used by all stakeholders involved in the early phases of a building renovation project if they have a certain knowledge about the project. It is aimed at the professional sector – i.e. not one-family housing and similar.

How?

In the development of the tool, the focus has been on making a tool that is easy to understand and simple to use. There are no new calculations needed to use the tool, and the interviews used for data collection are based on the existing facts about the project. An interviewer – for instance, a project assistant – can collect data from the different stakeholders with evaluations of the project on a scale from 1 to 5 for each of the eight parameters. The interview questions are standardised with possible minor adaptations depending on the building type.

Thus the evaluation of a project is based on subjective views, but it is supported by facts about the project. In addition, for each grade an interviewee gives, a reason should be stated. When the results are presented, who made the evaluations should be specified.

The evaluation should be repeated several times during the project and after it is finished.

Related tools

The tool can be used in combination with *strategies for sustainable* FM (see Chapter A.3).

Further information

The tool is presented in more detail in Chapter B.10.

A.16 Maturity model for strategic partnerships

What?

The tool is a maturity model that can be used to identify the maturity of an organisation's readiness to participate in a strategic partnership. The model consists of five steps, which represent five different ways to structure the relationship between the building client and the delivery team (consultants, contractors and possible material producers). The two axes, value and complexity, show that for each step, a more complex relationship is established, but also that a higher level of value is created. As an example, a relationship based on price is regarded as creating the least value, while a strategic partnership is expected to create most value.

Figure A16.1 shows that with increasing complexity in the relationship, the more it is possible to create value, but increased complexity also sets higher requirements for the different parties' abilities and maturity in relation to cross-organisational collaboration.

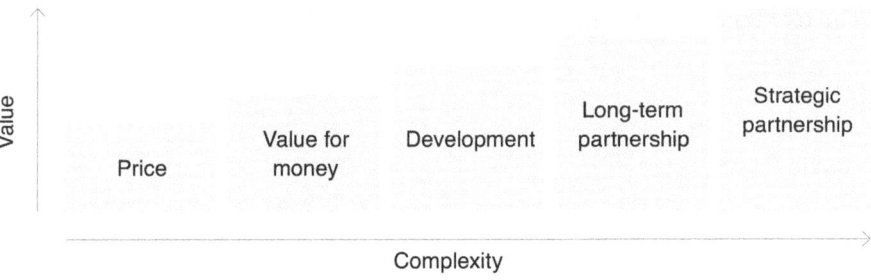

Figure A16.1 Strategic partnerships in construction maturity model

Why?

The purpose of the maturity model is to give the parties in the building industry at tool to analyse their own organisations' ways of operating and to identify areas where the companies can improve to be more prepared to participate in strategic partnerships. Experiences in the UK and Sweden show that there can be a number of benefits from moving one's building process to a higher level on the maturity staircase: Less conflict, higher budget certainty, higher user satisfaction and higher quality in building and renovation projects are just some of the benefits you can gain from a successful strategic partnership. However, not all companies or projects are suited to be part of a strategic partnership.

Who?

The maturity model can be used by and on all companies in the value and supply chain of the building industry (for instance, building clients, FM organisations, consultants, contractors and producers of building material) to evaluate their own maturity and the maturity of other potential collaboration partners. The focus can be on one's own company as part of a self-evaluation, but also on other companies that are potential collaboration partners and thus evaluated as part of a decision to eventually join forces in a partnership. Strategic partnerships are particularly relevant for collaborations that not only include one project but also a portfolio of building and/or renovation projects.

How?

You evaluate the maturity of a company by evaluating eight parameters for complexity and ten parameters for value (see Chapter B.20). All 18 parameters are evaluated on a scale of 1–5. The sum of these evaluations determines where the company is placed on the five steps of the maturity staircase.

Further information

The tool is presented in more detail in Chapter B.20.

A.17 From FM to building

What?

The tool consists of a typology of mechanisms to transfer knowledge from FM to a building project. The typology is based on principles for the transfer of knowledge:

- Pull/push
- Awareness/power
- Competences/codification
- Requirement/performance
- Control/responsibility
- Integration/outsourcing

The typology describes eight essential mechanisms that can be used independent of each other or in a combination – for instance, continuous briefing together with continuous commissioning.

Figure A17.1 shows on the left side four mechanisms for knowledge pull as requirements from FM and pulled from design. Similarly, on the right side,

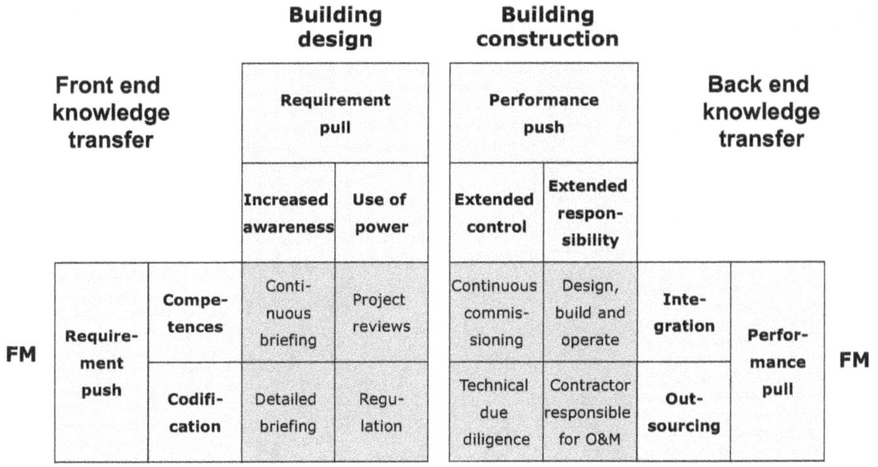

		Building design		Building construction			
Front end knowledge transfer		**Requirement pull**		**Performance push**			**Back end knowledge transfer**
		Increased awareness	Use of power	Extended control	Extended responsibility		
FM	**Requirement push**	Competences	Continuous briefing	Project reviews	Continuous commissioning	Design, build and operate	Integration	Performance pull	**FM**
		Codification	Detailed briefing	Regulation	Technical due diligence	Contractor responsible for O&M	Outsourcing		

Figure A17.1 Typology of knowledge transfer mechanisms from FM to building

knowledge transfer is pushed as performance validation from construction and pulled from FM.

Why?

The tool puts your focus on the transfer of knowledge from FM to building projects, and it helps you to consider specifically how you can ensure an efficient transfer of knowledge in the individual project and thereby possibly make the coming building more FM friendly.

Who?

The typology can be used by leading staff in FM and building client organisations as decision support and a basis for a dialogue between them in planning and organising building projects.

How?

You can use the tool for decision support to choose the right mechanisms to transfer knowledge. The typology shows which principles for transfer each mechanism utilises as well as which ones you can use in the design and the construction phases, respectively. You choose the mechanism(s) that are suitable for the specific project and integrate them, for instance, in the procurement strategy.

Related tools

The tool can be used in combination with *usability briefing* (see Chapter A.13).

Further information

The tool is presented in more detail in Chapter B.21.

A.V

Optimisation

Opening dialogue

Carl, FM team leader, to John, FM middle manager:

- I have a proposal for the work group, which we talked about forming for our next Lean project.

John's response:

- Fine. I will send an invitation to a kick-off meeting.

Why this process?

You should continuously ensure that the performance of the FM organisation is developing in the preferred direction and has the expected effect on the overall organisational performance of the corporation. Most FM organisations need to become more efficient and need to focus on how to do 'more for less'.

This process provides you with a systematic for working with your internal processes and procedures so that you use less time on 'firefighting' and get more time to do long-term planning.

What triggers starting this process?

- When a target for performance improvement has been defined by top management, for instance, as part of a general optimisation project
- When an initiative has been decided by FM management, for instance, because of unsatisfactory performance measurements
- When it is defined as a regular occurring task in the FM organisation
- When proposals from staff or collaboration partners are initiated

The process

The process for optimisation consists of six phases, which are described in Table A.10. The eight phases can be related the four phases of Plan, Do, Check and Act in the commonly used Deming quality circle, as shown in Figure AV.1.

Table A.10 Phases in the process for optimisation

Phase	Name	Description
A	**Evaluate current performance and improvement potential**	To conduct an optimisation process, it is essential that the FM organisation initially establish a baseline that the improvement can be evaluated against: What areas of FM are included and what is the starting point for the improvement? It can include both quantitative and qualitative measurements, and can, for instance, be based on satisfaction surveys among the users and benchmarking internally and externally. Based on this, you can estimate the potential of the improvements and define specific goals.
B	**Identify and decide changes**	It is highly important that you identify and decide on the right areas for improvement. This requires that you work creatively and not only focus on the 'usual suspects'. Therefore, it is recommended that you to arrange an idea-generating exercise among the involved staff. Another possibility is to exchange experiences and benchmark your internal processes with similar processes in other companies. Involvement of external consultants is also an often-used possibility. The decision to implement specific changes should be made by the management of the FM organisation.
C	**Implement changes**	To implement the specific changes, it is relevant to establish one or more project groups with the responsibility of realising each of their change processes.
D	**Evaluate new FM performance**	After the implementation is finished, an evaluation of the areas of performance that were included in the initial baseline should be made. You assess whether improvements have been made and whether the defined goals have been reached.
E	**Evaluate new organisational performance**	Similarly, an evaluation is made of the areas of organisational performance that were included in the initial baseline. Here you also assess whether improvements have been made and whether the defined goals have been reached. It would be appropriate to involve representatives from corporate management in this evaluation.
F	**Evaluate the need for further optimisation**	Based on the results of the aforementioned evaluations, you should consider whether there are possibilities to spread some of the changes to other parts of the company – for instance, other locations. It should also be considered whether there is a need to start further optimisation with other changes.

Tools for optimisation

Figure AV.1 shows the names of five tools that can be used for optimisation and the phases in the process where they are most relevant. A short description of each tool is given next, and this is followed by a section about each tool.

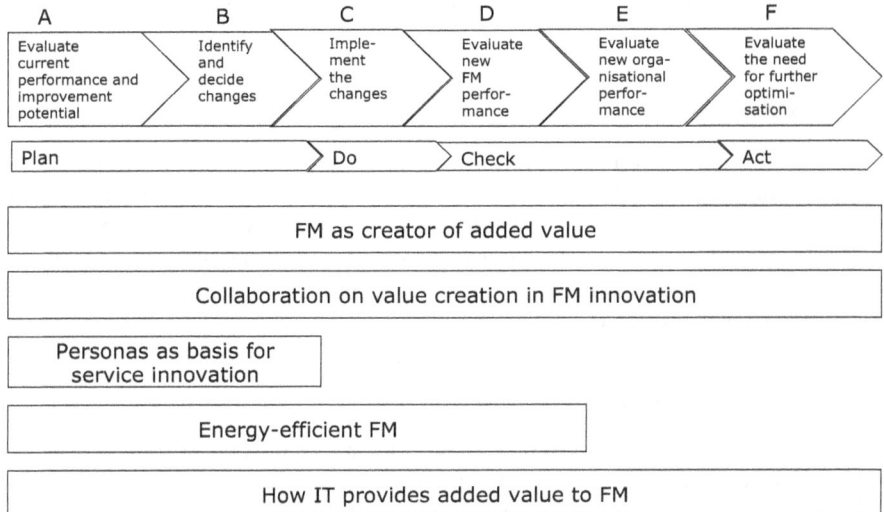

Figure AV.1 Tools for optimisation and process phases

FM as creator of added value

Among the most important challenges for facilities managers are initiating and managing changes which create added value for the core business and documenting the effects of the change to demonstrate that FM really does create added value. This tool consists of a simple process model combined with the general quality management model, Plan, Do, Check and Act. It distinguishes between output equivalent to FM performance and outcome equivalent with the overall organisational performance because only for the latter can we speak of FM creating added value.

Collaboration on value creation in FM innovation

It is a challenge to create innovation in FM because the demand side is very heterogeneous. Clients, customers and end users each have their specific needs and expectations. On the other side, this gives possibilities for providers to collaborate with different stakeholders to create value. This tool focusses on co-creation of value and presents a number of specific tools and methods for collaboration.

Personas as a basis for service innovation

Personas is a method that is often used in marketing to describe different target groups, but it can also be a basis for service innovation processes. This tool was applied in the context of service relationships in the non-profit housing sector, but it can be used more generally in FM organisations to obtain a better understanding of the end users. The tool is used in the initial phases A–B to evaluate current performance and improvement potential and to identify and decide changes.

Energy-efficient FM

This tool can help you in your work on optimising energy consumption. It consists of a maturity model which can be used as an analysis and decision support tool to arrange sourcing and eventual relationships with external collaboration partners through all phases of an optimisation process. The maturity model was developed with a focus on municipal FM organisations, but it can also give inspiration to other types of FM organisations.

How IT provides added value to FM

All FM organisations of a certain size need IT systems to support their processes, but not all implementation projects of IT systems in FM are successful. Generally, it is difficult to demonstrate the impact and thereby the value, of a new IT solution. This tool can help you with that. It can be used in all phases of an optimisation process and is divided in three steps. It includes, among other things, an analysis model to make an overall assessment of the efficiency and effectiveness of an IT solution.

A.18 FM as creator of added value

What?

The tool is based on a general process model with input → throughput → output → outcome:

- Input includes interventions or changes in FM which are implemented with the aim to create added value
- Throughput includes management of the implementation of the changes
- Output is the improvement in FM performance
- Outcome is the improvement in the performance and functioning of the core business, which is caused by the improvements in FM performance and therefore can be regarded as added value FM.

This is combined with the well-known Deming quality circle: Plan, Do, Check and Act.

Figure A18.1 shows the model that starts by defining input in the Plan-phase, continuous with throughput in the Do-phase, output and outcome in the Check-phase and finishes with evaluating in the Act-phase, which can lead to starting a new cycle. An extended version is included in Chapter B.29.

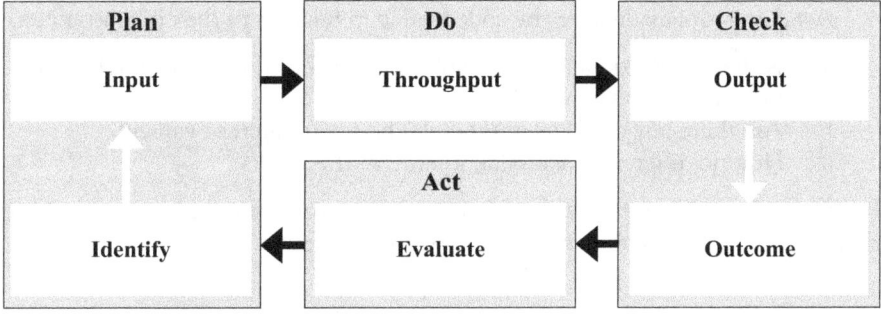

Figure A18.1 Model for Value-Adding Management

Why?

In recent years, in practice and research, there has been increased focus on how FM can create added value for organisations. In research, this has resulted in the development of a number of conceptual models and tools, as well as a collection of huge amounts of empirical information. However, the practical application of this knowledge has been limited and has shown to be difficult. The reasons seem to be that the different models are too complex and that there is a lack of common terminology and clear operationalisation of input-output-outcome relations. Therefore, we have developed a simpler model for the creation of a value-added model of FM.

Who?

Particularly, heads and managers of FM and project managers and developers will benefit from this tool, but much of the staff in FM organisations – in-house and external – can be involved, depending on the specific interventions. All decisions of major importance should be implemented by following a process that is planned in detail if the model is to be used as a management tool.

How?

The tool is based in the four steps included in the Plan, Do, Check and Act quality circle:

1   *Plan* starts by identifying the drivers for the change and defines the goals and condition to fulfil the goals. Then you define relevant interventions and decide which interventions should be *input* and therefore implemented.
2   *Do* is *throughput* in terms of management of the implementation and the related change process.
3   *Check* concerns the evaluation of *output* in terms of changes in FM offerings/performance and of *outcome* in terms of the performance and functionality of the core business. The improvements in organisational performance derived from the changes in FM performance are added value of FM.
4   *Act* consists of evaluating the added value in relation to the circumstances:

   a   Is the increased value reasonable compared to the cost of implementation and other sacrifices?
   b   Are there any conflicts or synergies between different values?
   c   How do different stakeholders perceive the results?

In addition, the FM strategy should be updated, and it should be considered if there is a need to initiate new improvement measures.

Related tools

The tool is of a general nature and can be combined with many other management tools and FM tools. It is a further development of the tools presented in Chapters B.25 and B.27.

Further information

The tool is presented in more detail in Chapter B.29.

A.19 Collaboration on value creation in FM innovation

What?

The method consists of an overview of tools that can be used to support the involvement of stakeholders in innovation processes and thereby contribute to organising and managing value creation in FM.

Table A.11 shows an overview of tools to support value co-creation in FM innovation.

Table A.11 Tools to support value co-creation

	User	*Resource*	*Co-creator*
Organisation as a whole/client	Ad hoc meetings	Transparency matrices and models	Regular and ad hoc meetings
		Workshops	Workshops
		Scenario analysis (with or without simulation IT)	
Internal FM unit/customer	Workshops	Workshops	Workshops
		Shared training	Face-to-face meetings
		Team-building activities	ICT for information management and sharing
		IT for information management and sharing	Team-building activities
			Scenario analysis
Employees/ end users	User surveys	User surveys	Shared training
	User workgroups	Face-to-face interviews	Idea competition
	Workshops	Workshops	Workshops
		Idea competitions	
		Shared training	
		Team-building activities	

Why?

An increasing understanding has emerged that companies no longer have full control to decide which values to offer the market. Instead, they have to continuously collaborate with their customers, who become active collaboration partners in value creation – in fact, value is created commonly between the supply-and-demand side. The former part offers the frame and resources to the collaboration on value creation, and the latter part makes their needs and expectations explicit, and shares their knowledge of how they can be satisfied.

The heterogeneity of the stakeholders involved in FM brings challenges in relation to the management of innovation. All the stakeholders' different needs and expectations should be considered when you have the responsibility of managing innovation and improvements in FM. Collaboration on the creation of value is a way for the supply side to manage innovation and improvements together with the demand side, but it requires an understanding of the complexity of FM.

Who?

The method can be used by all involved in the development of FM provisions – both in in-house functions and among external providers, as the understanding of complexity and relationships is important.

How?

The method can be used help to plan innovative processes and select specific tools and forms of collaboration. You should focus on your problem, where you stand, what you need to deliver and what is expected. The starting point is an analysis of the different stakeholders involved in the FM delivery on both the demand and supply side, and on the strategic, tactical and operational levels. It is important to understand the specific needs and expectations of different groups of stakeholders so that you can select the appropriate tools to support the collaboration between the relevant parties on each level.

Related tools

The tool can be used in combination with *drivers for innovation in services* (see Chapter A.4) and *personas as a basis for service innovation* (see Chapter A.20).

Further information

The tool is presented in more detail in Chapter B.16.

A.20 Personas as a basis for service innovation

What?

The tool presents examples of resident profiles in a housing association. These were developed by use of the method called personas, which is a specific way to work with target group analyses. A persona is a fictive description of a user, customer or, as in this context, a resident. The personas method was originally used in IT development but has spread to be used in marketing, communication, concept development and innovation. The basic idea is that if you talk to and understand your end users, you can develop persona profiles which will make it possible to improve your processes in such a way that the end users will (hopefully) be more satisfied.

Table A.12 shows three personas that were developed together with a Danish non-profit housing association because the association wanted to obtain more in-depth insight into their residents to be able to improve their service relationship. The profiles are described in detail in Chapter B.17.

Table A.12 Resident personas in a housing association

Person	Citations
The engaged, Hans	'I do not need to be trained in resident democracy; the housing association can just tell me what we are able to influence. Then I can deal with that, because we actually go into the area's committee because we want to make the place we live in a nice spot'.
The social, Inger	'Earlier, you just called the caretaker and then only after an hour he arrived. Sometimes, there was even time for him to sit down and drink a cup of coffee. This cannot happen today'.
The convenient, Camilla	'I can just call as soon as there is something that needs to get fixed. I think that's very nice, and it makes me think it is really nice to be a tenant'.

Why?

Personas support a common understanding of your customers. They can ascertain that decisions about creating new service offerings take as starting point, what the customers want and not what you and your colleagues think they want. Moreover, personas make it easier to remember who the customers are and to imagine their needs and motivations since they provide an understanding of the customers' different actions, perceptions and reception patterns. This is based on the assumption that, if you know their points of reference, you will achieve a better dialogue. Which is important when you are going to reorganise and optimise your procedures and consider how you create the maximum value for your users.

Who?

The tool is particularly relevant for team leaders, project managers and developers. However, everybody involved in the development of FM processes and in servicing users can benefit from using personas because they are key to knowing your beneficiaries in all processes.

How?

The method of developing personas can be applied to all users of FM and not just within the domain of housing associations. Unlike when working with segments, which mainly build on quantitative data, personas should be based on deep insight about the relevant target group, primarily based on qualitative data. The take on personas that is the most relevant depends on the individual type of FM organisation; therefore, each FM organisation should develop its own contextual persona.

When you use personas, it is important to be aware that a fictive element is involved so that the individual profile does not represent a portrait of a specific person. The purpose is to establish knowledge about the target group to be used in development, and therefore the individual profile should be seen as representing the sum of knowledge collected across interviews and document data. You can base the development work on all the personas so that the whole user group and its often contradicting needs are addressed, or you can focus on a single persona if the aim is to develop more target-oriented communication.

The FM staff can develop personas themselves, or you can acquire assistance from external specialists. For each person profiled, you should fill in the following:

1   Name
2   Age
3   Profession
4   Family situation

5   Hobbies
6   The person profiled appreciates that . . .
7   The person profiled hates that . . .
8   Other questions relevant to your users and organisation

It is important that you make a visual expression – e.g. printed on large posters or carton figures placed across the whole organisation. In that way, personas, and hence the perspective of the end user, become part of ongoing innovation processes.

Related tools

The tool can be used in combination with *drivers for innovation in services* (see Chapter A.4) and *collaboration on value creation in FM innovation* (see Chapter A.19).

Further information

The tool is presented in more detail in Chapter B.17.

A.21 Energy-efficient FM

What?

The method consists of a maturity model for how a municipal FM organisation can be assessed and improved in terms of energy-efficient FM. It shows that external competences and collaboration with private partners can contribute to energy efficiency and other performance criteria.

Every step on the maturity staircase represents a phase with an increasing value of energy efficiency. Each phase includes some key processes, and the municipal FM organisation is assessed by a descriptive yardstick for each of them. The exercise results in a maturity level ranging from 'ignorant' to 'professional'. Awareness of the current situation and improvement potential should support decisions on which initiatives should be taken in-house and which should be taken in collaboration with external service providers specialising in energy efficiency.

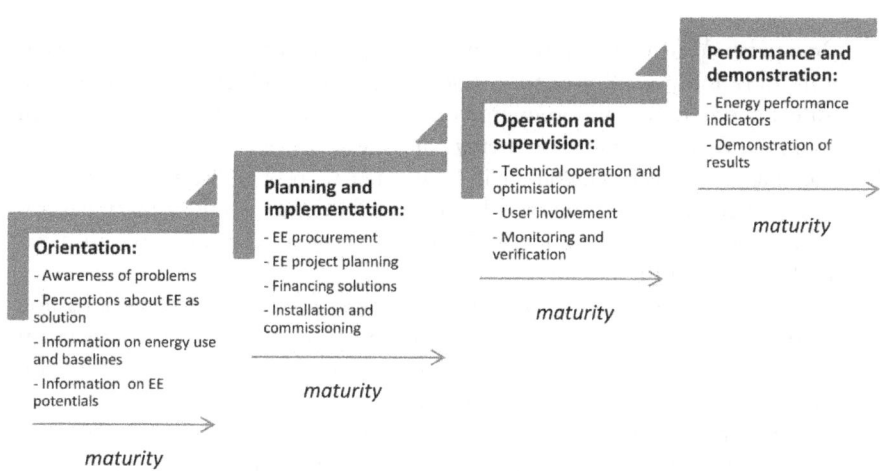

Figure A21.1 The maturity model consisting of four steps with increasing maturity

Why?

It is important for most FM organisations to become more energy-efficient, and it is not necessarily the best solution to have all competences in-house. The method provides a tool to assess the internal capabilities and to analyse advantages and disadvantages between different solutions and levels of ambition. If you have the capacity and competences, it might be most appropriate to implement the improvements in-house. It is in any case important that the improvements are made stepwise. Every new step is dependent on the earlier one and is a precondition for the following step. The organisation needs time to adapt. A maturity approach can in principle be developed and used to optimise anything related to daily operations – for instance, cleaning, learning among staff members, etc. It contributes to work with the development of the FM function in a systematic way.

Who?

This type of collaboration will often be project oriented and involve contractual partnerships, but the strategy for the actual partnership needs to be planned and structured in-house. The head of FM, the project manager and the team leader are central and need to provide information from the daily operation. The method was developed in relation to municipal FM organisations, but it can also be used by other FM organisations.

How?

It is important to analyse needs, internal competences and benefits from the activities before possible collaboration with external partners is started. The method includes the following four steps:

1   *Orientation.* You create awareness about the problem and investigate the precondition and possibilities to develop more energy-efficient solutions. You collect the needed data on energy consumption and estimate the potential savings and other benefits of energy efficiency.
2   *Planning and implementation.* If it is decided to be beneficial, you arrange procurement to engage external partners. The energy-efficiency measures are planned, and the financial solution is clarified. The measures are then implemented in the buildings.
3   *Operation and monitoring.* You carry out ongoing technical operations and energy optimisation. Along the way, you involve the users with a focus on the behavioural aspects of energy consumption. The energy performance is measured and monitored continuously.
4   *Demonstration of energy savings and other benefits.* You use indicators to demonstrate the saving in energy consumption during the period starting the energy-efficiency measures and over time during the continuous energy

optimisation. In addition, you document other benefits – for instance, in terms of improved indoor climate with reduced draught, more stable temperatures and improved air quality.

Related tools

The tool can be used in combination with *capacity building in FM organisations* (see Chapter A.7) and *ESCO as a method for learning in FM organisations* (see Chapter B.9).

Further information

The tool is presented in more detail in Chapter B.11.

A.22 How IT provides added value to FM

What?

The tool is a general method to assess, how IT provides added value to FM. The method builds on an understanding of added value, which involves a fundamental distinction between efficiency (internal resource optimisation) and effectiveness (external effect optimisation). Another important distinction is between use value and economic value (exchange value).

Figure A22.1 shows an example of a conclusive assessment of IT creating added value, which involves an assessment of whether the IT solution at the same time results in a high degree of efficiency and a high degree of effectiveness. If both is a reality, then the IT solution provides added value.

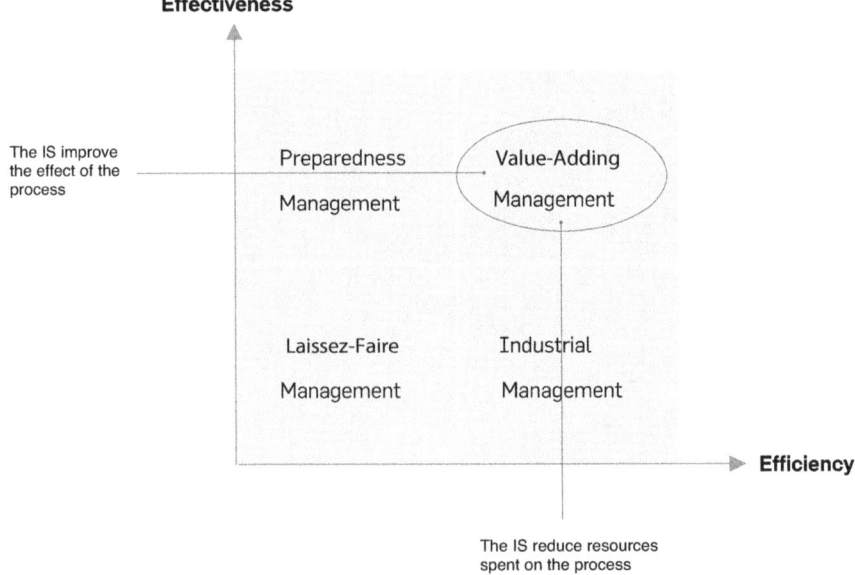

Figure A22.1 Assessment of efficiency and effectiveness of IT in FM

Why?

It is a major challenge to assess the added value of initiatives related to implementations of IT to support processes in FM. It is often unclear what added value is expected and which part of the value chain in FM benefits from IT. This is the background for developing this general method to assess the added value of IT in FM.

Who?

The tool is particularly suitable to be used by project managers and developers involved in IT projects in FM organisations, but the principles in the tool can also be used on other areas.

How?

The tool is based on a combination of a number of concepts, such as the FM supply chain, IT functionality and Value-Adding Management. It includes the following three steps:

1   Analyse the IT solution by using a model of the FM supply chain. Identify the activities, roles and levels involved in the use of the IT system.
2   Assess the increase in efficiency and in effectiveness, as well as the functionalities of the IT solution.
3   Assess to which degree the IT solution contributes to Value-Adding Management and whether added value is created.

Related tools

The tool is partly based on the tool *Value-Adding Management* (see Chapter B.27).

Further information

The tool is presented in more detail in Chapter B.28.

Part B
Models, methods and tools

Introduction to part B

The overall research profile for CFM was from the beginning formulated as follows:

Research in
space for people,
buildings with use value and
property and infrastructure that facilitates

The core of CFM's activities has been research projects. These have been distributed on five main themes, as shown in Figure B.1.

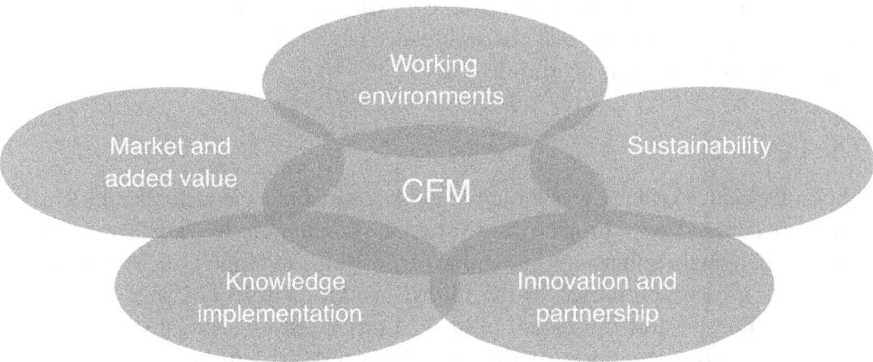

Figure B.1 Themes for CFM's research

These themes are also the basis for the structure in Part B, which is divided into the following six sub-parts:

  I Concerns working environments under the heading 'Facilities that Support Users and Activities'
 II Concerns sustainability under the heading 'Sustainability from Goals to Action'

III   Concerns innovation and partnerships, which is also the heading
IV   Concerns knowledge implementation under the heading 'Transfer of Knowledge from FM to Building Projects'
V   Concerns market and added value under the headline 'FM and Added Value'
VI   Concerns FM more generally and has the headline 'FM Organisation and Development'

Each sub-part includes between four and eight chapters, and each chapter concerns a project or results, including a central illustration of a model, method and/ or tool, and possibly short descriptions. The criteria for selection and presentation was that the results should be applicable in practice, but it, of course, varies both for whom they are relevant to be used by and how easy they are to use in practice.

To evaluate how the different models, methods and tools can be used in practice, we have in each chapter utilised the same five typical processes that form the basis for Part A, where the processes are described in detail. These evaluations have, as mentioned in the preface, been made by the editor – in some cases together with the author.

Literature references in the text have been kept to a minimum for easy readability. Instead, there is a guide to literature at the end of each chapter, which provides references to the publications the chapter is based on and possibly an overview of other publications in English from the same research project and/or author. This provides a basis for finding more information about the topic of the chapter. Many of the publications can be downloaded from www.cfm.dtu.dk and/ or the publication database at the university where the researchers are or were employed. At DTU, this is orbit.dtu.dk.

The future of FM

The last chapter of Part B – Chapter B.34 – concerns scenarios for FM in the future. It is based on a project we finished approximately five years ago. It ends with my personal evaluations based on the possible developments in two dimensions with great uncertainty for the future development in FM: the economy in Northern Europe and the global climate. I described five scenarios and named them after road-metaphors. At the time, I expected the future of FM would be closest to the scenario in the middle called 'The Roundabout', where there were neither particular positive nor negative prospects for both dimensions.

How has the development then been in the meantime, and how do I now evaluate the future of FM based on the same dimensions and scenarios? For the economy in Northern Europe, the development has as a whole been fairly positive, but without signs of long-lasting high growth. On the contrary, Brexit and the other economic and political problems in the European Union, and the risk of new housing and other financial bubbles are signs that a high-growth scenario is not just around the corner. Concerning the global climate, the Paris Agreement in 2015 was a clear positive signal, but the election of Trump as the American president has definitely weakened this, and the increasing global conflict level in relation to Russia, Iran, China and North Korea is threatening in relation to a united effort against climate change.

Altogether, I, therefore, evaluate that we are moving towards moderate economic growth in Northern Europe and a negative development in relation to the global climate. Thus we are coming closer to the scenario I called 'The Winding Road', which is characterised by the following:

- Less growth in globalisation
- Strong focus on Corporate Social Responsibility (CSR)
- Focus on national development and renovation due to lack of resources
- Relocation of parts of cities from low-lying areas due to rising sea levels
- Reduction in use of building space due to rising site prices and building cost
- Health, safety, security and environment in focus

For FM, this means that the handling of local and physical resources will become increasingly more important. However, it is always difficult to make predictions – particularly about the future!

B.I

Facilities that support users and activities

B.1 Typology of office workplaces

Juriaan van Meel

Introduction

The development of office workplaces is relatively young as a research field. Much of the present-day understandings and theories are based on ideas from the 1990s, when pioneers like Francis Duffy from the planning company DEGW in the UK and Franklin Becker from Cornell University in the US started to describe and analyse new office types and workplaces. Since then, a tremendous amount of literature on new workplace concepts has emerged. Within this literature, there is a striking gap between scientific publications and the more popular, visionary books on office layout.

The research project on knowledge workplaces, which Juriaan has conducted for CFM, aimed to bridge this gap. Like the popular literature, the purpose was to identify the comprehensive changes in the work environment, but unlike the literature, empirical data and an analytical framework were applied to describe and understand these changes. Another main difference was that the study did not focus on the future. It was a fundamental idea that the relevant challenges and concepts already existed in some countries and companies. Thus the study has tried to identify these cases and describe, discuss and analyse them from different angles – not only with a focus on design and technology but also on their context, the underlying development process and their strategic value.

The purpose of the research project, therefore, was to describe the current state of knowledge workplaces. The project has provided a re-evaluation of 30–40 years of experimentations, research and debate on workplace concepts, such as remote working, open offices and flexible workplaces. Among the questions were the following: What is the outcome of 20 years of experience with 'alternative offices' and 'new ways of working' (NWOW)? How has it affected the working environments around the world in relation to mobility, space utilisation and quality of workplaces? And what can we expect? Are there new trends in the design of workplace concepts emerging which challenge the traditional perception of the office?

Besides describing knowledge workplaces, the study focussed on analysing the processes that create and shape these working environments. What are the most important drivers – and barriers – for changes? Who are the stakeholders who dominate the design of workplaces? And how does it affect the outcome of the development process? By answering these questions, the project has contributed to a more profound understanding of knowledge workplaces and shed new light on the possible future development.

The study has mainly applied two methods:

- Historical analysis: description of the development of the working environments since around 1970 and until now based on literature review and interviews
- Case studies: selection, description and comparison of case studies, which provides an accurate picture of present and possible future working environments around the world

The typology

The result of the project is a typology of workplaces. The typology was presented in the book *Workplaces Today* (Van Meel, 2015). In Figure B1.1, the typology is presented with graphical icons for each of the ten types and a short description of each type.

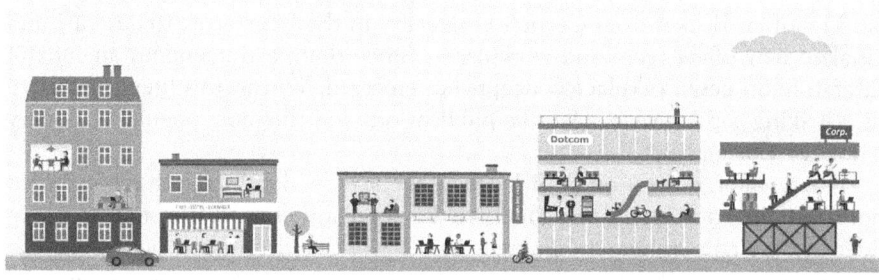

TYPE 1: HOME OFFICES	TYPE 2: PUBLIC SPACES	TYPE 3: CO-WORK OFFICES	TYPE 4: PLAY OFFICES	TYPE 5: FLEX OFFICES
Large numbers of people work from the comfort of their homes: at the kitchen table, in the study or on the sofa. It is a popular option, appealing to many. But working from home also means working alone. There is an ongoing discussion about how this impacts the social cohesion of organisations.	People can be found texting, typing and talking in cafés, libraries and parks. These spaces are not designed for work, but they are appropriated as such by mobile workers, students, ordinary office workers and freelancers. Every place is a workplace?	Co-working means working on your own, but alongside others in a shared office space. Co-working protagonists point out that the concept is not only about sharing space, but just as much about sharing a sense of community, thereby creating possibilities for synergy and collaboration.	The play office is the office as playground, with bright colours, slides and foosball tables. Tech companies in particular seem to be fond of such cheerful environments. Are these offices whimsical gimmicks or a welcome deviation from the bland efficiency of traditional office design?	In flex offices, office workers no longer have their own desks. Instead, they share a variety of 'activity-based settings'. It is a space-saving concept that makes a lot of sense from an economic and sustainability point of view. But it is not always easy to persuade people to give up 'their' office territory.

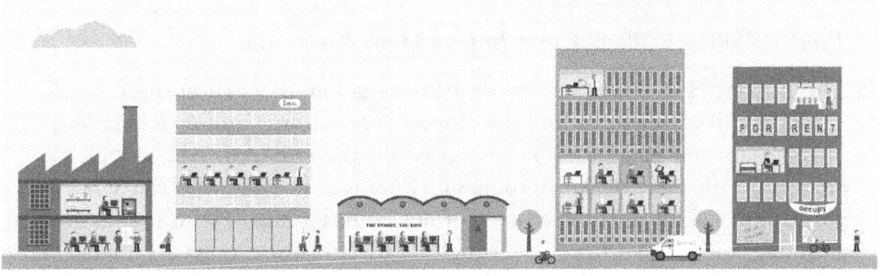

TYPE 6: *STUDIOS*	TYPE 7: *MODERNIST OFFICES*	TYPE 8: *PROCESS OFFICES*	TYPE 9: *CELL OFFICES*	TYPE 10: *RECYCLED OFFICES*
Studios are spaces for creative work, where tables hold not only computers but also models, samples, sketches, books and other creative artefacts. Studios are interesting because they are designed to foster creativity, collaboration and learning – critical qualities for any organisation.	The modernist office is the classic glazed box with neutrally coloured interiors filled with large numbers of identical desks. These offices are designed as rational machines for working in without any ambition to be cosy, playful or trendy. They are true places for work.	Process offices are offices where information is processed rather than produced. Think of crowded call centres and other places for 'low-end' office work. Such places are normally absent from books like this, but they are an essential part of today's digitalised economy.	Cell offices are associated with long corridors and rows of rooms. This type of office is close to extinction because it is considered incompatible with today's collaborative working ethos. Some companies, however, cling to it, seeking to create a calm environment for cognitively demanding work.	The office has been declared dead many times. Let's just assume those predictions do indeed come to pass. What then to do with all those empty office buildings? The most productive answer is 'adaptive reuse', giving former office buildings a new life as apartments, hotels or healthcare facilities.

Figure B1.1 Typology of office workplaces

Cases

The book presents a total of 30 cases – three cases for each of the 10 types. They are distributed with 18 from Europe, 7 from the US, 2 from China and 1 from Australia, Hong Kong and Mexico, respectively. The 18 cases from Europe included 6 from the Netherlands, 3 from Denmark, 3 from the UK and 6 from other countries. Three specific examples of cases are presented briefly next.

Type 3: Co-work office. Case: Impact Hub, Amsterdam

The Impact Hub is a place for idealists targeted at social entrepreneurs. There are designers, product developers, consultants and coaches. Most of them are self-employed. It is located on the site of a redeveloped gas factory on the edge of Amsterdam's city centre. The co-work office occupies the first floor of the factory's former administration building. Spaces are large and filled with light. Much of the furniture is second-hand or self-made. The Impact Hub in Amsterdam was established in 2013, and it is part of a wide network of Impact Hubs across the world. The first Impact Hub (originally just called 'The Hub') was established in London in 2005. Since then, the Impact Hub has grown into an international association with more than 50 locations on six continents and over 7,000 members.

Type 4: Play office. Case: Lego, Billund

At the Lego's headquarters in Billund, Denmark, a new building was established in 2010 for Lego's department for product development. The building has 2,000 m² divided on two floors with a large light atrium. From the upper to the lower floor, you can go down through a tubular slide if you do not prefer to take the stairs. The office is a base for more than 130 designers and engineers who come from all over the world. Most workplaces are open, and much of the work involves transdisciplinary project groups. There are open as well as enclosed meeting areas with bright colours. Lego models are placed everywhere – on work surfaces, tables for model building, cupboards and shelves.

Type 7: Modernist office. Case: GlaxoSmithKline, Philadelphia

The global healthcare company GlaxoSmithKline (GSK) was one of the early adopters of flexible office concepts. For almost a decade, the company has been experimenting with open and shared workspaces, improving the concept with each new project. Yet GSK's new office in Philadelphia, US, is the company's first building designed and built entirely around the concept of 'SMART working', as it is known in GSK parlance. The objective of the new work environment from 2013 was to increase both the efficiency and effectiveness of GSK's work processes. GSK knew that its old offices were used less than 35% of the time they were available. Creating a more intensively used, shared environment, therefore, resulted quite easily in significant savings on the lease and on operational expenses.

Supplementary information about the research project

The original title of Juriaan's research project for CFM was Knowledge Workplaces in Europe. The project started in 2009 with a pilot study aiming at more specifically defining the research project. The main project took place in two phases. The first phase from 2010 to 2012 focussed on uncovering the historical development of office workplaces. This included a comprehensive study of both scientific literature and the more popular, professional literature since the end of the 1960s. One of the results was a scientific paper (Van Meel, 2011) which identified the first ideas, initiatives and investigations concerning new ways of working (NWOW) back in the 1970s. One of the striking early examples was the 'Mobile Office' by the Austrian architect Hans Hollein, who sat and worked with a drawing board and telephone in a glass bubble, which he also called 'The Portable Office in a Suitcase'. Another result was a book chapter with six contrasting cases of office layout from the Netherlands (Van Meel, 2012). Juriaan, together with two co-authors, worked on transferring a Dutch publication they had written into a new international book: *Planning Office Space* (Van Meel, Martens og Van Ree, 2010). This book is now published in English, French, Chinese and Spanish. CFM supported the work on the book to a limited extent, and we had a special version published with CFM's logo for promotional purposes.

The original plan was that the research project should investigate office workplaces based on cases from different European countries – including the Netherlands and Denmark. When the proposal for the second phase was presented to CFM's centre committee towards the end of 2012, it was recommended that we include cases from other parts of the world. The book *Workplaces Today* (Van Meel, 2015) was published simultaneously in English and French. The English version is published in collaboration with ICOP and CFM. Both mentioned books are available at amazon.com.

Literature guide

The typology is presented in full in the following book:

van Meel, Juriaan: *Workplaces Today*. ICOP and Centre for Facilities Management – Realdania Research, Technical University of Denmark, 2015.

Six contrasting cases from the Netherlands are presented in the following book chapter:

van Meel, Juriaan: Office contrasts: Six contrasting Dutch cases. Chapter 3.2 in Per Anker Jensen and Susanne Balslev Nielsen (eds.): *Facilities Management Research in the Nordic Countries: Past, Present and Future*. Centre for Facilities Management – Realdania Research, DTU Management Engineering, and Polyteknisk Forlag, January 2012.

The origin of new ways of working with new office concepts in the 1970s is presented in the following scientific journal paper:

van Meel, Juriaan: The origins of new ways of working: Office concepts in the 1970s. *Facilities*, Vol. 29, No. 9/10, 2011, pp. 357–367.

A general guide on planning office spaces is presented in the following book:

van Meel: Juriaan, Yuri Martens and Hermen Jan van Ree: *Planning Office Spaces: A Practical Guide for Managers and Designers*. Laurence King Publisher. A version with special cover published by CFM for promotion, Autumn 2010.

Other publications from CFM with Juriaan van Meel as sole or first author include the following:

van Meel, Juriaan: Workhubs. *Research Design Connections*, No. 1, 2011.
Van Meel, Juriaan and Rikke Brinkø: Working apart together. *EuroFM Insight*, No. 30, September 2014.
van Meel, Juriaan and Rikke Brinkø: Working apart together. FM World: The BIFM Facilities Management magazine. *British Institute of Facilities Management*, 27 February 2014.
van Meel, Juriaan and Mikkel Thomassen: Using narratives for design briefings. *Metropolis Magazine*, November 2012.
van Meel, Juriaan and Hermen Jan van Ree: Workplace design made accessible. *Office et Culture*, January 2011.

Publications from CFM with Juriaan van Meel as co-author include the following:

Brinkø, Rikke, Susanne Balslev Nielsen and Juriaan van Meel: Access over ownership: A typology of shared space. *Facilities*, Vol. 33, No. 11/12, 2015, pp. 736–751.
Brinkø, Rikke, Juriaan van Meel and Susanne Balslev Nielsen: The shared building portfolio: An exploration and typology. *Proceedings of CIB FM Conference, Technical University of Denmark, 21–23 May 2014*.
Van Ree, Herman Jan and Juriaan van Meel: Space exploration. *FM World*, January 2011, pp. 24–27.

The use of the typology – assessed by Per Anker Jensen

As shown in the table that follows, the typology of workplaces is, in particular, useful in two of the five processes mentioned in the introduction to the book: strategy development and space planning. For both processes, the typology can be used as a tool for terminology and analysis, possibly in combination with creativity and facilities (Chapter 2) and how to share space (Chapter 5). The described cases can also be used for inspiration.

Process	Phase							
Strategy development	A	B	C	D	E	F		
Organisational design	A	B	C	D	E	F		
Space planning	A	B	C	D	E	F	G	H
Building project	A	B	C	D	E	F	G	H
Process optimisation	A	B	C	D	E	F		

For strategy development, the typology can be used by FM management and staff at the strategic level as a tool for terminology and analysis in the internal work on mapping the status for the company's current workplaces, definition of strategy goals, development of strategy plans, implementation, follow-up and adjusting plans (phases B+C+D+E+F). In addition, the typology can be used as a basis for dialogue with top management of the company, managers of other departments/business units and external collaboration partners, both in the process of defining strategy goals and developing strategy plans (phases B+C). Moreover, the typology can be used in further work as a tool for terminology in the dissemination of the strategy to the other staff in the FM organisation and staff members who will be affected in other departments during implementation and follow-up (phases D+E).

For space planning, the typology can be used by staff on a tactical level as a tool for terminology and analysis in the internal work on mapping status for the company's current supply and demand of spaces of different categories, to identify and evaluate alternative space allocations and to re-evaluate space utilisation (phases A+B+C+H). In addition, the typology can be used as a basis for dialogue with FM management and the affected departments about decisions (phase D) and with the affected departments and external collaboration partners during planning, implementation and evaluation of space reconfigurations (phases E+F+G).

B.2 Creativity and facilities

Birgitte Hoffmann

Introduction

How can one support the development of a creative environment? What is the significance of physical facilities? And what is a creative environment all together? These are some of the main questions which have guided CFM's research project on facilities for creative environments. The project takes as starting point that activities of creativity no longer exclusively belongs to specific departments of development and innovation in businesses or at dedicated organisations like incubators and science parks. Today, most public and private businesses and organisations are supposed to address creativity as part of their practices. To explore how creativity can be supported, an interdisciplinary literature review supplemented with six case studies of very different facilities has been conducted, and multiple perspectives on creative environments have been analysed and described.

The discourse on creativity has had a huge influence on the development of cities, private and public companies and society as such during the last decades. Creativity is seen as a key to a better future by a number of actors, including the United Nations, which emphasises that the creative economy has the potential to generate growth and at the same time strengthen social inclusion, cultural diversity and human development. It seems like a major 'wave of creativity' has poured over us.

The global fascination with creativity must be seen as related to the general reorganisation of western societies on the backdrop of the financial crises, energy crises, technological development, institutional changes and globalisation in general. Thereby, creativity as a concept has become a driving force, not only in the 'creative industries' but also for the management of public institutions and private companies in general. These organisations have a strong focus on strengthening creative environments with the aim of utilising the energy expected to be in this wave. Thus new strategies and re-development projects are forwarded in many organisations in the name of creativity, and there are naturally many who want to get on the creativity wave rather than risk being left behind.

Hence, there are good reasons to take a closer look at the wave and those surfing on it, to stay with the metaphor. This chapter presents results from an analysis of six very different case studies of 'creative environments' in Denmark and how these are being established and facilitated – both by physical facilities and by processual facilitation, including (re)organisations. We analyse the perceptions of creativity and of the facilitation of creativity as they are expressed in the cases, as well as how these perceptions have developed through the practical 'use' of the environment. The case studies are based on a theoretical and historical analysis of the development of the concept 'creativity' and the emergence of the current creativity discourse.

The analysis is explorative; hence, we try to show the diversity, both in the perceptions and the implementation of these concepts with the aim of making a more thorough reflection on how creative environments are interpreted and worked with. Thus the reader should not expect a definition of creative environments or detailed models for establishing or strengthening them. It is precisely a main point that it is not possible to work with creative environments in such a way. It takes more, and this 'more' is what we try to present some pictures of and entries to. In that way, we want to inspire and contribute to creating insights and discussions, and thereby strengthen the creative environments.

Facilitating creative environments

The discourse on creativity has many long roots, but it really came on the agenda in the western world around the turn of the millennium with the concept of 'the creative class'. This put a focus on creativity and creative environments as essential growth drivers for cities in the post-industrial world. Hence, where earlier many cities planned to strengthen a 'business climate', they today are more preoccupied with planning what, based on the American regional planner Richard Florida, can be named a 'people climate'.

A parallel narrative on creativity can be derived from a managerial perspective with a series of theoretical and practical approaches. Hence 'organisational learning' has been a headline in the development of organisations in the knowledge society that draws learning and personal development from formal settings to become an integrated part of the lives of organisations and individuals. Several approaches explicitly dealing with 'creativity' can be identified in the fields of design and innovation.

From the different contexts, the idea of creativity as a main driver of social processes of change has fused, and inspiration and learning are diffused across different environments. Cities, and in particular the lively city environments coined in the concept of 'life between buildings' by the Danish architect Jan Gehl, are clear sources of inspiration for the development of companies that integrate cafés, 'ramblas' and artwork as a frame for informal meetings and inspiration from what's different. Cities have at the same time adopted a strategic development approach from the business world and work with branding, benchmarks, strategic city renewals and engage with private organisations to get a share of the development.

Six exploratory case studies

To explore how creativity is conceptualised and performed, six very different cases are analysed:

Mindlab is a concrete and spectacular case of how the global focus on innovation and creativity is unfolded in a central public administration. Mindlab was established as an in-house facility in the Ministries of Industry, Businesses and Finance to develop the innovative competences of public workplaces with the explicit criterion of developing societal welfare. The case illustrates that the vision of creativity in practice needs to navigate in the cross-pressure between welfare and an efficiency discourse.

SCION DTU is organised in relation to the Technical University of Denmark and is a more traditional example of how entrepreneurialism is supported by creating a frame for individuals and small groups to support them in developing 'an idea'. Creativity is focussed on business development by specific, concrete facilities and some networking. In this way, SCION DTU tries to shape or discipline creative people into market realisation.

The retrofitting of *Tryg*, a large insurance company, is directly related to the discursive 'need' to be an innovative and attractive work environment. In a radical redesign of both the buildings and the organisation, Tryg supports a rather wide understanding of a creative environment relating communication, collaboration and work environment. They draw heavily on elements from the 'creative city' about informal meetings and diversity of spaces. However, the case portrays a dilemma between unfolding the informal urban culture of the 'rambla' and the formal business principles with, for example, strict rules of 'clean desks'.

The case of *Musicon* directly relates to 'the creative city' discourse as the municipality tries to redevelop a former industrial area to support the development of Roskilde as an attractive city to live and work in for the 'creative class'. Still, the concept of creativity is linked to traditional fields and perceptions of art as an especially creative domain. By implying this interpretation of creativity, the municipality may miss the opportunity of cross-fertilisation between the cultural sector and innovative activities and businesses.

Herning is a provincial municipality and an example of how planners and politicians seize the concept of creativity and adopt the idea to support local growth. In this case, existing local narratives of art and entrepreneurialism were rewritten to support the municipal identity and development. While young subcultures often are in focus in the development of creative environments in the larger cities, in Herning, the activities are performed by informal networks across established public and private organisations and businesses.

The Candy Factory is an activist cultural environment that has developed in a former candy factory in agreement with the owner in the outskirts of Copenhagen. The Candy Factory is an example of the new type of businesses that are called for as a basis for urban life and growth. During the development of the candy factory, a formalisation of the organisation as well as a development of a strategic perspective shows the concrete and ongoing negotiations of creativity.

Photos of the cases are shown in Figure B2.1. Table B2.1 presents a summary of the six cases based on their relations to the analytical headlines: facilities, facilitation, use and culture.

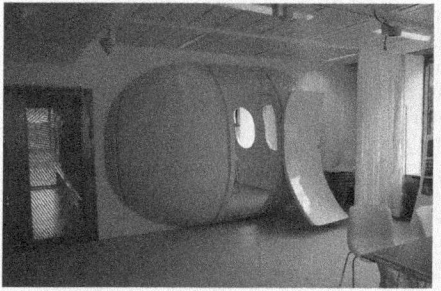

Mindlab

Tryg Insurance

SCION DTU

Musicon

Herning

The Candy Factory

Figure B2.1 Photos of the six cases

Table B2.1 Summary and comparison of six cases of creative environments

	Facilities	Facilitation	Use	Culture
Mindlab Innovation of welfare services Life between departments	Designed facilities 'The mind' as a symbolic stage Flexible rooms, easy to retrofit	Designed processes Professional facilitators Customised processes 'User' focus	Departments apply for support Mind lap as a 'neutral zone' Focus on facilitation – not all facilities are used	'Staged informality' 'Out of the box' User journeys as the outset
Tryg Insurance Future workplace Life between the offices	Designed Urban ideal Open offices, workshop squares, art, 'ramblas', and cafés	Designed, CEO-led, participation, company rituals, monitoring, storytelling	Generally well received Critique of open offices Varied use of 'ramblas'	Wide/diffuse focus on creativity Formal business: 'clean desk'
SCION DTU Creative entrepreneurs Life between businesses	Designed Private offices Shared meeting facilities plus services and recreational facilities	Make it easy to spend more time on business development Networking activities	Less use of recreational facilities than the demand for them Vertical (tenant-landlord) relations more than horizontal (networking)	Businesses development in focus Autonomous units – with opportunities for common learning
Musicon 'The creative city' Life between the buildings	Planned 'non-design' Retrofitting large and raw industrial buildings to new urban area	Design 'Plan as little as possible and as much as necessary'	Little informal life Mostly well organised invited functions	Narrow perception of creative culture Formal planning Business growth
Herning 'Regional competition' Life at trade fairs	Art Museums Huge events Trade fairs	Bottom up Local networks Support private initiatives	Mostly users from out of town Landmarks/symbols Revenue	Liberal Pragmatic David and Goliath We can do it Self-made
The Candy Factory Culture revolution, fun spirit Life	'Non-design' Non-standard sculpture in itself Former industrial buildings used for reflection, production, exhibitions, parties	Diverse activities Non-script Common meetings User groups Use	Simultaneous production of culture and space	Informal Activist Self-realisation Structures sneaking in

The cases mirror a diversity of facilities and approaches to facilities and facilitation. In the case of Mindlab, one the one hand, the focus is on the facilities, not least because a very spectacular meeting facility developed for Mindlab has a strong symbolic value. On the other hand, the facilitation is the core of Mindlab, as the focus during its use has turned away from the spectacular facilities and the creation of ideas to the facilitation of long-term processes aiming at integrating the novelties in practice.

At SCION DTU, the focus is on the basic facilities, and the recreation facilities have mainly symbolic value. The facilitation is considered supplementary, but SCION DTU did experience a need for supporting social processes and therefore employed a person with this specific function.

In the case of Tryg, the need to modernise the facilities initiated the development, but what started out as a physical endeavour turned out to include many organisational efforts to change the culture according to the ideas behind the design.

In Musicon, the 'raw facilities' are the basic potential: open for redesign and use. The facilitation in the form of urban development processes aims at opening up the area for the creative new developers and to support the urban life in and between the buildings. The 'creativity' is supposed to come along with the new inhabitants, and thus the facilitation only relates to the urban development that in practice enacts a gatekeeper function. This case depicts a dilemma between openness and uncertainty versus formal planning practices and political control. Thus this facilitation may accidentally hamper the development as it narrows down the 'creativity' to be included.

In Herning, the focus is also on more formal art and cultural events. What 50 years ago was very spectacular art supported by a visionary art-loving local business man has now turned into a large number of art museums and cultural events, as well a local narrative of being an 'art city'. Not because the locals have turned into art lovers themselves, but because 'they can do it', a David and Goliath narrative rebellious to what others may think in relation to the city's size and location in the mid-west of Denmark. The municipal planners use the concept of creativity to support the local narrative of trade and entrepreneurialism; however, they also to try to challenge local understandings of urban life – e.g. the dominating car city.

The Candy Factory is at the heart of the 'creative city', even though the participants explicitly try to distance the organisation from this discourse. Urban planners and developers think of their presence and activities as a way to create urban liveability and in this way raising prices for the plot. In this way, the political activist environment may contribute to processes of gentrification.

Conclusions

We have established an analytical framework based on the theoretical studies and the case studies, which can be used as a starting point for reflection and discussions of development of creative environments. We have formulated and discussed a number of interesting perspectives in relation to creative environments and the relation between facilities and facilitation and organisational culture.

The project does not support a focus on facilities as a key strategy to strengthen creative environments, which otherwise are manifested in a number of special designs of rooms and offices. The physical facilities by themselves will rarely be enough to strengthen a creative environment, only the physical facilities may have roles as symbolic markers. Instead, one must take a broader perspective on the processes and the cultures of the organisation.

The analysis points to competent facilitation as essential to change work practices and relationships; hence, facilitation can help the development of creativity in environments that do not traditionally work creatively. Along these lines, several of the cases showed a development where the facilitation became more important than intended in the planning. Altogether, the cases paint a picture of a progression in the understanding of creativity from a focus on the 'good ideas' towards innovative processes and learning as complex, long-lasting and integrated processes that need support in relation to developing existing practices.

In a continuation of this, the analysis emphasises the importance of the underlying 'culture'. The culture prevailing in the local company or the planning culture of a municipality is crucial for the ability and willingness to support creative work and creative environments. Without a fundamental openness and support to the creative work, it seems difficult to develop and sustain creative environments.

This means that the creative environments function in a complex interplay between the physical facilities, facilitation and culture. The constellation of these elements has a completely different local expression, and the cases indicate that even if there are general models, local anchoring and continuous development is crucial.

The cases point to a number of current challenges. Establishing creative environments challenge the existing practices like, for instance, the formal structures and time horizons of planning, as well as the division in sectors and thinking in silos. At the same time, the open and informal are under pressure in several places from other discourses, which prioritises brands, politics and economic efficiency. It should also be emphasised that research in stress indicates that the modern creative workplace develops new forms of stress and breakdowns. Those working with supporting creative environments need to address these challenges; otherwise, these cross-pressures may hamper the facilitation of creative environments.

Literature guide

The overall results of the research project are documented in a final report in Danish:

Hoffmann, Birgitte, Morten Elle, Peter Munthe-Kaas and Jan Lilliendahl Larsen: *Kreative Miljøer – mellem faciliteter og facilitering*. Research Report. Centre for Facilities Management – Realdania Research, DTU Management Engineering. November 2012.

Publications in English from the project include the following scientific article:

Larsen, Jan Lilliendahl, Morten Elle, Birgitte Hoffmann and Peter Munthe-Kaas: Urbanising facilities management: The challenges in a creative age. *Facilities*, Vol. 29, No. 1/2, 2011, pp. 80–92.

In addition, the author has published two book chapters:

Hoffmann, Birgitte, Morten Elle and Peter Munthe-Kaas: Facilitating creative environments: Lessons from Danish cases at different organizational scales. Chapter 3.4 in Per Anker Jensen and Susanne Balslev Nielsen (eds.): *Facilities Management Research in the Nordic Countries: Past, Present and Future*. Centre for Facilities Management – Realdania Research, DTU Management Engineering, and Polyteknisk Forlag, January 2012.
Hoffmann, Birgitte, Peter Munthe-Kaas and Morten Elle: Facilitating creative environments. Chapter 5 in K. Alexander and I. Price (eds.): *Managing Organizational Ecologies: Space, Management and Organization*. Routledge, 2012.

Also, two conference papers:

Hoffmann, Birgitte, Morten Elle and Peter Munthe-Kaas: Facilitating creative environments: Lessons from Danish cases at different organizational scales. CFM's Nordic FM Conference, 22–23 August 2011.
Hoffmann, Birgitte, Peter Munthe-Kaas, Jan Lilliendahl Larsen and Morten Elle: Facilitating creative environments: When the winds of creativity hit FM. *EFMC2010, Madrid, 1–2 June 2010.*

Use of the framework – evaluated by Per Anker Jensen

Creativity and facilities seen as a tool is mainly a conceptual framework. As shown in the table that follows, the framework is primarily useful in three of the five processes presented in Part A: strategy development, space planning and process optimisation. The framework can in all processes be used as a terminology and analysis tool, eventually together with typology of workplaces presented in Chapter B.1 and how to share space presented in Chapter B.5. In addition, the described cases can be used as inspiration.

Process	Phase							
Strategy development	A	B	C	D	E	F		
Organisational design	A	B	C	D	E	F		
Space planning	A	B	C	D	E	F	G	H
Building project	A	B	C	D	E	F	G	H
Process optimisation	A	B	C	D	E	F		

For strategy development, the framework can be used by FM management and staff on the strategic levels as a terminology and analysis tool in the internal work on mapping the status of the company's current surroundings, definition of strategy goals and development of strategic plans (phases B + C + D). In addition, the framework can be used as a basis for dialogue with corporate top management, managers from other departments/business units and external collaboration partners, both in the process of defining strategy goals and the development of strategic plans (phases B+C).

In space planning, the framework can be used by staff on a tactical level as a terminology and analysis tool in the internal work on mapping the status for the company's current supply and demand of space of different categories and identify and evaluate alternative space plans (phases A+B+C). In addition, the framework can be used as a basis for dialogue with FM management and departments affected by decisions (phase D).

For process optimisation, the framework can be used by FM management and staff on the strategic level to identify and decide interventions (phase B).

B.3 Usability briefing

Aneta Fronczek-Munter

Introduction

This chapter presents a model for usability briefing. The purpose of the model is to provide an easily understandable overview of the activities of usability briefing, defined as a dynamic and continuous process of capturing user perspectives throughout all the phases of building projects. The starting point for the development of the model was the design of hospital buildings, but the model can also be used for other complex buildings.

Usability is a concept similar to functionality that is mainly used in the development of user interfaces for IT software or mobile phones. For such products, usability tests are often arranged during product development. This typically takes place in a laboratory, where potential users are invited to test prototypes to evaluate whether the products are intuitively understandable and easy to operate. During the last 15 years, particularly in the Nordic countries in Europe, there has been development in the application of the usability concept to buildings and workplaces in use.

In the PhD study, which this chapter is based on, we tried to take this a step further by focussing on how the results of such evaluations can be used in the briefing of new building projects. It provides knowledge about capturing user needs and defines a process model for usability briefing for hospital architecture from a user perspective. The research results generate a better understanding of how knowledge about user needs from workshops or evaluations can be fed into briefing and design processes.

The model

The model is shown in Figure B3.1 and includes the following four activities: (1) briefing, (2) evaluation, (3) user involvement and (4) design. The horizontal time axis in the figure is divided according to the phases in a typical building project, including pre-project with phases 0 and 1, project with phases 2–5 and post-project with phases 6 and 7. The phases are based on the often-used international Plan of Work 2013 from the Royal Institute of British Architects, but it also resembles the commonly used division of phases in Denmark.

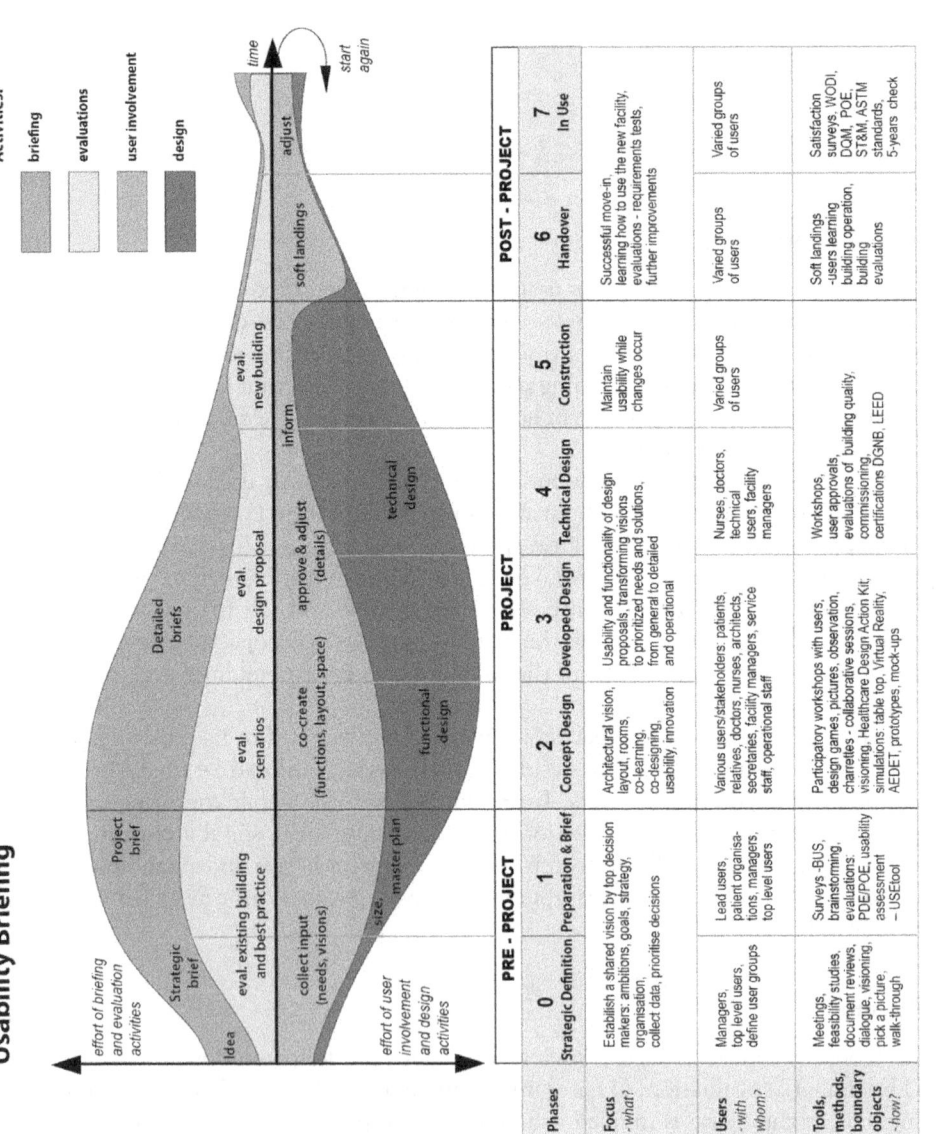

Figure B3.1 Model for usability briefing

The four activities are seen as essential in a proper briefing process with a focus on usability. They are shown as separate, parallel layers with varying intensity along the time axis, but the activities should interact with each other during the different phases. The model is based on the precondition that the briefing does not finish with a complete building brief in an initial phase but goes on as a continuous process with varying focus and intensity during the whole project duration. It is also a precondition that the briefing to a high degree takes place as a facilitated and mutual dialogue between users and designers, where co-creation in terms of co-learning and co-design are important elements.

The table underneath the model in Figure B3.1 shows recommendations with regard to specifying focus (what?), involved users/stakeholders (who?) and tools, methods and boundary objects (how?) in the different phases. Boundary objects are artefacts, documents, illustrations and ideas that can function as shared means of communication for people with different backgrounds, such as users and designers. Examples of boundary objects in a briefing and design process are scenario descriptions of buildings to be designed, reference photos, building briefs, drawings, physical models of building or floor plans, BIM models or other visualisations or mock-ups of (parts of) buildings.

What is new in the model is that it combines the four known activities, briefing, evaluation, user involvement and design, and arranges them through the whole project duration in relation to each other, and it recommends the most important focus areas, user groups and methods that can be used in the individual phases. The model puts awareness on the early phases and on usability across activities and time.

A precondition for the model is that topics related to usability are formalised in each of the four activities (for instance, in agendas, notes and documents) discussed (for instance, in workshops, design meetings) and evaluated in a systematic way (for instance, in project reviews). Furthermore, it is a precondition that the topics are kept explicitly in focus when a change from one activity to another takes place so that topics are not lost along the way. Thus the focus on usability should be continued throughout all phases. The model has been developed with the intention of being generic and easy to adapt to be used in the planning of new complex building projects with the inclusion of the activities relevant for the specific project.

Case of Bispebjerg Hospital

The PhD project included three case studies of hospital buildings: Bispebjerg Hospital in Copenhagen, Healthcare Innovation Lab at Herlev Hospital near Copenhagen, and Saint Olav's Hospital in Trondheim. In the following, only the case on Bispebjerg Hospital (BH) is described briefly as an example. The case was based on interviews and participation in internal workshops, etc., in 2010–2012 during the preparation of the brief for the master plan competition. The BH case focussed on phase 1 in the model, which is a part of the pre-project and covers preparation and brief. In this phase, briefings, evaluations and user involvement has very high intensity. The purpose is to develop a building brief that can form a basis for starting the actual project.

In BH, they appointed six thematic user groups, and all carried out a series of three workshops with each having its on specific focus. The goals were ambitious, but the groups were aware that their influence was limited, so the members knew that not all their wishes would be fulfilled. In fact, a central part of their task was to identify dilemmas which needed to be balanced in the further planning of the project. On example of this was that the final building brief concerning the plan for traffic and parking presented the following dilemma: 'Overview of the site and easy orientation should be accomplished. How can that be achieved while the historic identity of the site with small intimate areas and niches is sustained at the same time?'

BH had good experiences with using workshops with lead users – i.e. selected staff members with long and broad experience at the hospital. This group, for instance, prepared an overall model for the structure of the hospital with proximity relationships between all the hospital functions. To ensure continuity, BH also appointed a continuous user group to follow the project through all phases.

Specific tools mentioned in the model

The model and the table in Figure B3.1 mention specific tools, including many with abbreviations. The PhD project produced an overview and a categorisation of these and a huge number of other tools. The following is an overview of some specific tools mentioned in the model:

AEDET (Achieving Excellence Design Evaluation Toolkit): British methodology for evaluating design of hospital buildings

ASTM standard (American Society for Standards and Testing): see ST&M

BUS (Building Use Studies): British method for staff satisfaction surveys

DQM (Design Quality Method): British method to evaluate the design quality of buildings

Healthcare Design Action Kit: American checklist to be used by managers and architects and methodology for patient investigation concerning buildings in use

PDE (Pre-Design Evaluation): Methods for investigation before a rebuilding project

POE (Post Occupancy Evaluation): Methods for evaluation of buildings after occupation

Soft landings: British methodology to ensure a smooth transition from building project to use and operation

ST&M (Serviceability Tools and Methods): American system to evaluate whether requirements concerning the serviceability of buildings are fulfilled. ST&M has been developed to an ASTM standard

USEtool: Norwegian methodology for evaluation of the usability of facilities

WODI: Dutch system for staff satisfaction surveys and benchmarking

Supplementary information on the research project

The title of Aneta's PhD project was 'Usability Briefing of Hospital Facilities; Exploring User Needs and Experiences to Improve Complex Buildings'. Besides usability, evaluation and user involvement, the project also focussed on innovation – particular with regard to user-driven innovation. Her first conference paper concerns an investigation of the relation between usability and user-driven innovation (Fronczek-Munter 2011a). The case study on Healthcare Innovation Lab at Herlev Hospital also had user-driven innovation as a starting point, and Aneta has treated this case as an example of user-driven innovation in another conference paper and a book chapter (Fronczek-Munter, 2011b and 2012). The case study on Saint Olav's Hospital in Trondheim included an evaluation of part of the renewed hospital complex by application of the Norwegian methodology USEtool for evaluation of usability of facilities. Further information about USEtool can be found in Chapter B.4.

Literature guide

The model for usability briefing is described fully in Aneta's PhD thesis:

Fronczek-Munter, Aneta: Usability briefing for hospital design: Exploring user needs and experiences to improve complex buildings. PhD Dissertation. Department of Management Engineering, Technical University of Denmark. April 2016.

A more concentrated overview can be found in a recent conference paper:

Fronczek-Munter, Aneta: Usability briefing for hospital architecture: Exploring user needs and experiences to improve complex buildings. *European Healthcare Design Conference, London, 11–14 June 2017.*

Overviews of specific tools can be found in the thesis and in the conference papers that follow:

Fronczek-Munter, Aneta: Evaluation methods of hospital facilities. *International Journal of Facilities Management*, 2013, pp. 216–227. EuroFM.
Fronczek-Munter, Aneta, Van der Zwart, Johan, Hansen, Geir Karsten: How to evaluate healthcare buildings? Selection of methods for evaluating hospital architectural quality and usability: *A case at St. Olavs Hospital in Norway. ARCH17 Conference, Copenhagen, 26–27 April 2017.*

Aneta's publications also include the following:

Fronczek-Munter, Aneta: Usability and user driven innovation: Unity or clash. 13th International FM&REM-Congress: "Built Environment", Kufstein, Austria, 27–29 January 2011a.
Fronczek-Munter, Aneta: Facilitating user driven innovation: A study of methods and tools at Herlev Hospital. CFM's Nordic FM Conference, 22–23 August 2011b.
Fronczek-Munter, Aneta: Facilitating user driven innovation: A study of methods and tools at Herlev Hospital. Chapter 5.5 in Per Anker Jensen and Susanne Balslev Nielsen (eds.): *Facilities Management Research in the Nordic Countries: Past, Present and Future.* Centre for Facilities Management – Realdania Research, DTU Management Engineering, and Polyteknisk Forlag, January 2012.

Fronczek-Munter, Aneta: Usability briefing: A process model for healthcare facilities. *Proceedings of CIB FM Conference, Technical University of Denmark, 21–23 May 2014.*

Jensen, Per Anker, Keith Alexander and Aneta Fronczek-Munter: Towards an agenda for user-oriented research in the built environment. *Proceedings from the 6th Nordic Conference for Construction Economics and Organisation, Copenhagen, April 2011.*

Use of the model – assessed by Per Anker Jensen

The model for usability briefing is, as shown in the table that follows, particularly useful in two of the five processes presented in Part A: building project and process optimisation. The model can in both processes be used as a general planning and management tool, including overview of specific tools, possibly in combination together with value-based space optimisation (Chapter B.4). The described cases can also be used as inspiration.

Process	Phase							
Strategy development	A	B	C	D	E	F		
Organisational design	A	B	C	D	E	F		
Space planning	A	B	C	D	E	F	G	H
Building project	A	B	C	D	E	F	G	H
Process optimisation	A	B	C	D	E	F		

The model is primarily targeted to be used in building projects. It can be used by managers and staff on the strategic level in client and FM functions as a planning and management tool and as a basis for dialogue with stakeholders in building projects concerning arranging briefing, evaluation and user-involvement activities, including decisions on which specific tools to apply. This is in particular relevant in the initial phases during pre-project but should also be part of the evaluations during post-project (phases A+B+H). On the tactical level, the model can be used in the implementation of the mentioned activities and as a basis for dialogue with users and external collaboration partners during the whole duration (phases A–H).

For process optimisation, the model is primarily useful in processes involving changing of buildings, including new layouts of spaces and workplaces. Here the model can be used by FM management and staff on the strategic level in the evaluation of current performance and improvement potential, identify and decide changes, evaluate new FM performance and new organisational performance and evaluate the need for further optimisation (phases A+B+D+E+F). For staff on tactical level, the model can, in particular, be used in implementation of changes and as a basis for dialogue with users and external collaboration partners (phase C).

B.4 Value-based space optimisation

Mette Tinsfeldt

According to a Danish FM handbook, approximately 20% of the totals costs in a company are related to operation and maintenance of the buildings, and the same applies to higher educational institutions (HEI's). Thus, a quantitative space optimisation can have a significant effect on the overall economy of both a company and a HEI.

However, it is essential to acknowledge that a quantitative space optimisation does not necessarily ensure the optimal physical environment for the activities the space is supposed to support. In the case where a quantitative space optimisation impairs the conditions for a company's work processes, a potential decrease of the employee's productivity can eat up the expected cost reduction, which the space optimisation initially was supposed to realise. Therefore, a space optimisation should include a quantitative as well as a qualitative aspect to ensure that the optimisation realises the intended added value.

Denmark's HEI's are facing several challenges. These are among others related to an adjustment of the number of students in each class and the fact that the HEI's in 2007 went from being state owned to self-governing. Another challenge is that their buildings and spaces aren't designed to support the present way of teaching, which to a higher degree facilitates group work rather than traditional blackboard teaching. Space optimisation can help overcome some of these challenges if it is based on a holistic analysis approach and is conducted with the acknowledgement that space optimisation not only concerns placing more people on less space but also has an influence on the teaching environment as a whole.

The present project initially examined which elements should be included in a space optimisation process when applied at a HEI in order to contribute the most possible added value. Furthermore, the project concerned a space optimisation process at two HEI's based on the evaluations' methods Post Occupancy Evaluation (POE) and USEtool. The space analysis was based on a significant user involvement as a mean of getting insight in how the institutions' primary stakeholders – namely, the students and the teachers – experienced the institutions present spaces and furthermore get their ideas and wishes in relation to a space optimisation.

Based on the results from the two evaluation methods, and the students' and teachers' statements, the project identified several space optimisation initiatives, which advantageously could be implemented based on the current conditions at the respective institutions. Lastly, the project analysed the evaluations methods POE and USEtool in relation to the conducted space analysis, which resulted in a suggestion of how a space optimisation could be conducted in order to realise added value for HEI's in general.

The suggested space optimisation analysis process consists of seven phases, which include walk-throughs, interviews, hand out of questionnaires and workshops with the users of the actual spaces subject to the optimisation (see Table B4.1).

Table B4.1 Space optimisation process with seven phases

Phase	Objectives	Activities
1	Clarification of purpose and success criteria	Interview the head of school and the facilities manager
	Identifying stakeholders	Walk-through with the head of school and the facilities manager
	Preparing project plan and clarify resources	
	Collection of data about the organisation, buildings and space challenges	
2	Collection of data about use of space	Observations and interviews
		Analyse space utilisation
3	Discussions about the existing use of space	Focus groups with the primary stakeholders
4	Clarifying which space solutions work well and which do not – generally and related to specific aspects of the analysis	Walk-through with the primary stakeholders
5	Involvement of larger groups of stakeholders	Questionnaire survey
6	Preparation of proposals for space optimisation and implementation plan	Workshop with the primary stakeholders, the head of school and the facilities manager
7	Implementation of space optimisations	Churns, rebuilding, etc.

Case studies

The case studies included two gymnasiums in the Greater Copenhagen area. Herlev Gymnasium is located in a suburb in Herlev Municipality north-west of Copenhagen and consists of a single building from 1976. The building is a typical concrete construction, two stories and rectangular. The gymnasium has approximately 800 students and 110 employees.

Falkonergården, the second gymnasium, is located in an older urban area in Frederiksberg Municipality encircled by Copenhagen Municipality. The gymnasium consists of several connected buildings built in 1955, and it currently rents additional spaces nearby in order to accommodate the space demand. Falkonergården has approximately 1,000 students and 100 employees.

The case studies concluded that none of the two gymnasiums had succeeded in integrating their spaces in their overall strategic goals for the institution. The analysis confirmed that it is essential, when conducting a space optimisation, that it is based on the values and strategic goals for the organisation, At Herlev Gymnasium, the objective with the space optimisation was mainly qualitative with a great focus on creating spaces, which supports the teaching to a higher extent, motivates the students as well as the teachers and attracts more students. At Falkonergården, the main objective was quantitative, focussing on utilising the existing spaces to a higher degree in order to create room for an increasing number of students and furthermore reduce the need for external spaces for teaching.

Besides defining the objective and clarifying the concrete prerequisites at the two institutions, it is also important to include an analysis of the less-specific aspects such as the culture and the users' habits, as this can show to be essential for the specific space optimisation proposals. A holistic approach can ensure that the space optimisation proposals are not only based on a superficial and theoretical understanding but also actually create added value when implemented. Both aspects can be identified by user involvement in the process, which, furthermore, can ensure that optimisation of one area is not conducted on the expense of another.

The project resulted in 11 specific proposals for space optimisation at Herlev Gymnasium and 7 at Falkonergården. Though the objectives of the space optimisations and the layout at the two buildings were very different, several striking similarities were detected:

- Lack of development of the schools' buildings corresponding to the changes the teaching activities have undergone, including a higher degree of group work
- Limited and insufficient spaces for the teachers to work and prepare the teaching
- Limited possibilities for conducting quantitative space optimisation

The project cannot conclude whether the aforementioned challenges are generic for all gymnasiums in Denmark; however, we assume that several others, which also are located in older buildings, have challenges equivalent to the ones Herlev Gymnasium and Falkonergården are facing.

Literature guide

The chapter is based on a conference paper:

Tinsfeldt, Mette and Per Anker Jensen: Value adding space management in higher education. *Proceedings of CIB FM Conference, Technical University of Denmark, 21–23 May 2014.*

Use of the method – assessed by Per Anker Jensen

As shown in the table that follows, value-based space optimisation is particularly suitable to be used in two of the five processes presented in Part A: space planning and process optimisation. The method can in both processes be used as a planning tool, possibly in combination with the method for how to share space in Chapter B.5. In addition, the described cases can be used for inspiration. The method was developed for gymnasium buildings, but it can also be used for other types of buildings, perhaps with minor adaptations.

Process	Phase							
Strategy development	A	B	C	D	E	F		
Organisational design	A	B	C	D	E	F		
Space planning	A	B	C	D	E	F	G	H
Building project	A	B	C	D	E	F	G	H
Process optimisation	A	B	C	D	E	F		

The method is, in particular, aimed at processes with space planning and can especially be used by staff on the tactical level with responsibility for space management in FM functions and consultants involved in similar tasks. This involves all phases from evaluation of the demand and supply of space to re-evaluation (phases A–H).

For process optimisation, the method is mainly suitable for processes that include space optimisation. Also, here the method can be used by staff on a tactical level with responsibility for space management in FM functions and consultants involved in similar tasks. This involves all phases from evaluation of current performance and improvement potential to evaluation of the need for further improvements (phases A–F).

B.5 How to share space

Rikke Brinkø Berg (born Rikke Brinkø)

Introduction

The use of shared space as a real estate and space management strategy, especially outside the field of office and workspace, is a relatively new area within both practice and research. The shared space mind-set in itself though, sharing our spaces and facilities with each other, is not a new idea. Sharing a multitude of different aspects of our lives, such as summer homes and public transport, is such an integrated part of our daily lives that we do not even consider them as *sharing* in any special way anymore. In addition to these examples, shared space has been used as a design method in the field of traffic planning and road design for many years.

Many topics have contributed to the development of shared space as a concept within the field of the built environment. Inspiration and knowledge can be found within the more established areas of FM – for example, in connection with the design of office spaces, but also from the fields of urban planning; traffic planning, as already mentioned; and more. Another field that has played a significant role in the development of shared space is the sharing economy, which promotes a mind-set of access over ownership and has grown explosively over the past decade, pushed forward by new technology and social media.

As we move away from these more well-described fields onto sharing with several different organisations involved and in many other contexts than we are used to consider sharing in, then a large gap emerges in our knowledge base. A knowledge gap ranging from which factors are the most important from a user point of view to how one as an owner or manager can maximise the chances for success, and which challenges and benefits can be expected, just to mention some.

The PhD project 'Sharing Space in the Knowledge City' was launched to fill this gap and to contribute with new knowledge about shared space in the context of FM in order to lay the foundation for shared space to be used as a new method for a more efficient and sustainable operation of buildings and properties. The project was divided in two main phases. The first focussed on exploring how shared space can be defined in a FM context, as well as what types of shared spaces exist, how to categorise these and create a language with which shared space can be discussed and developed. The second phase focussed on identifying essential aspects of the complex processes involved in working with and in a shared space, and how the new knowledge developed could be operationalised and contribute to the realisation of shared spaces in practice.

To answer these two questions, four main methods has been applied during the project: literature reviews, case studies, interviews and workshops. The end result is twofold: a typology of shared use of spaces and facilities, and a guide on how to create a shared space in a municipal building portfolio. Both will be presented in the following section.

Typology of shared use of space and facilities

The typology of shared use of space and facilities is one of the main results from the PhD project. The thought behind the development of the typology was that the information contained in the typology could form the basis for a more detailed discussion about the different aspects of shared space – for example, in order to clarify wishes and needs during a process of creating a new or developing an existing shared space. The typology is focused on the sharing of facilities, spaces and buildings between users from different organisational contexts and divides shared spaces into three different degrees. These degrees are determined by the level of interaction and collaboration required at an organisational level, ranging from invited sharing – the least involved, across cooperative sharing – to complete sharing – the one requiring the highest level of inter-organisational collaboration (see Figure B5.1).

Following the general division of shared spaces into the three degrees is a table containing a number of keywords used to describe essential characteristics of the

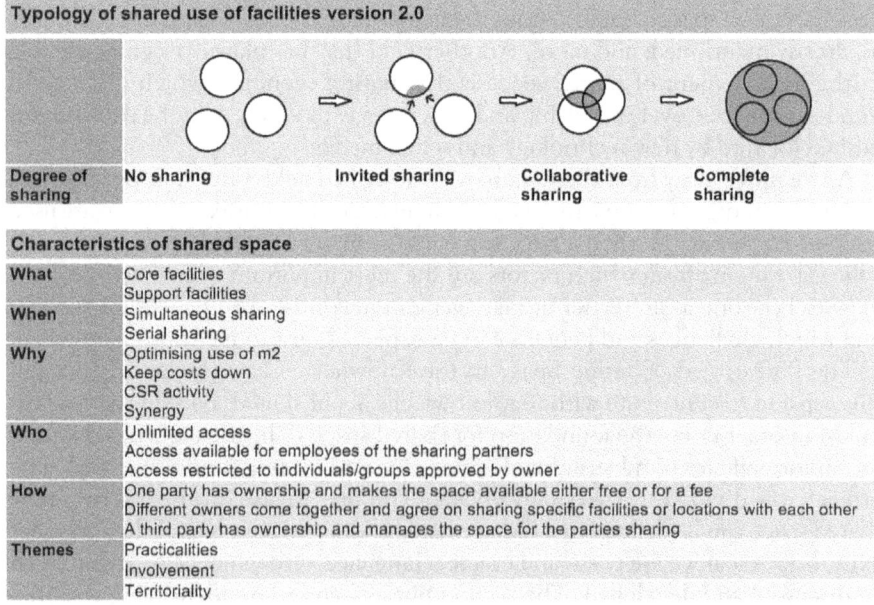

Typology of shared use of facilities version 2.0

| Degree of sharing | No sharing | Invited sharing | Collaborative sharing | Complete sharing |

Characteristics of shared space		
What	Core facilities Support facilities
When	Simultaneous sharing Serial sharing
Why	Optimising use of m2 Keep costs down CSR activity Synergy
Who	Unlimited access Access available for employees of the sharing partners Access restricted to individuals/groups approved by owner
How	One party has ownership and makes the space available either free or for a fee Different owners come together and agree on sharing specific facilities or locations with each other A third party has ownership and manages the space for the parties sharing
Themes	Practicalities Involvement Territoriality

Figure B5.1 Typology for shared space

specific shared space being worked on – either theoretically/hypothetically or physically. These are *what* (what is shared), *when* (when is it shared – simultaneously or sequentially), *why* (the motivation/purpose of the sharing), *who* (who is involved in the sharing) and last, but not least, *how* (how is the sharing organised). As mentioned earlier, these keywords should be used to develop a thorough description of the shared space to form a solid base for further discussions and development as well as the final realisation of the shared space in practice.

In addition to these keywords, the table also contains three additional aspects to be considered in the process of establishing or managing a shared space, and these are *territoriality*, *involvement* and *practicalities*. During the PhD project, these three themes have been identified as essential when working with or in a shared space, and they all have to do with different aspects of interaction between users and the daily operation of shared spaces.

The first of the three themes, *territoriality*, can be seen as the biggest barrier to shared space and represents the problems that can occur when people who are used to having their own space suddenly have to share with others. A number of aspects play a role in relation to this theme, such as knowledge of the party the space or facility will be shared with and co-determinations or participation in the process. The next theme, *involvement*, can be seen as the most important tool in overcoming the problems that can occur, such as for example are territoriality. As the name implies, it mainly deals with involving the parties that will be sharing a given space in the process. The last theme, *practicalities*, concerns the many different logistical and organisational aspects and considerations which must be dealt with in connection with working with and in a shared space. Everything from safety to cleaning.

Cases

The typology is developed based on a large number of cases, and in the following, a case representing each of the different degrees of sharing will be presented.

Invited sharing: Microsoft, DK

Microsoft Lyngby is Microsoft's new domicile in Denmark. The building project ran from 2013 to 2015 and was initiated to replace Microsoft's two existing locations in Northern Zealand. The new location has in addition to the private work areas that make up most of the building, two main areas that will be shared with parties from outside the organisation. The first is a number of workspaces that students from the local area can apply for, and these will be available for approved students during regular office hours. The students will have access to the first part of the building, which is not open to the public, while the rest of the buildings still will be off limits. The second main area is a public café on the ground floor level. Due to security considerations, the café, originally planned as one unit, was divided in two. One section to serve Microsoft employees and one to serve the general public through a separate entrance in the façade.

Collaborative sharing: Zeeburgereiland, NL

IKC Zeeburgereiland is a new educational facility in Amsterdam, Holland, and was completed in 2013. The building include space for a nursery, a kindergarten, primary school (<12 years), after school care, a sports arena and a number of unspecified 'functions for the local community' – all in one building. The project was developed to tackle a number of societal challenges, among which were providing sufficient facilities to a fast growing newly developed area with rising population numbers. Due to a stricter legislation regarding safety, hygiene, etc., regarding the smallest children, the nursery has its own space within the building as well as a separate entrance. The remaining partners have separate 'home zones' within the building, which are private, but outside these zones, the rest of the building and its functions are shared. The sport facility is used by all, and the after-school care is located in existing classrooms after school is over, just to mention a couple of examples of sharing. In addition to the physical aspects, most of the services needed to run the building are shared, such as cleaning, catering etc.

Complete sharing: Lyngby Idrætsby, DK

Lyngby Idrætsby is a non-profit sports facility in Denmark, owned by Lyngby-Taarbæk Municipality. The construction project, which included an extensive renovation and expansion of the existing facility, ran from 2012 to 2016, and increased the size of the complex from approximately 13,700 m², not counting outdoor areas, to 23,080 m². This increase made it possible to add a number of facilities to the existing complex, including new spaces for the sport associations, an area reserved for the business community, a day-care facility and space for the Lyngby-Taarbæk Youth School, all in order to support the municipality's vision of creating an area that is to be characterised by activity for as many hours a day as possible, for as many different users as possible. The facilities, both old and new, are to be shared where possible, and all new facilities are planned with multipurpose use in mind, with much emphasis being put on flexible interior and décor in order to achieve this goal.

A Guide to Shared Space in Municipalities

The second main result of the PhD project is a guide to establishing a shared space in a municipal real estate portfolio. The guide was developed in collaboration with representatives from municipalities and the private sector, and represents the complete work and knowledge produced during the three-year PhD project. The cover of the guide is shown in Figure B5.2.

Figure B5.2 A Guide to Shared Space – cover of version in English

The guide is divided into two main sections with two different focus areas. The first part provides a general introduction to the subject shared space, focusing on what shared space can contribute with, how to identify buildings and spaces with potential in relation to establishing a shared space and what challenges can arise during the process. The second part contains a step-by-step 'manual' for establishing a shared space in a municipal real estate portfolio.

The guide is constructed as a stand-alone product and contains the manual for how it is to be used and will therefore not be presented here in further detail.

The guide can be downloaded via the website www.cfm.dtu.dk in an English version as well as a Danish version.

Literature guide

The typology and the guide are described and discussed in full detail in English in Rikke's PhD thesis (please note that Rikke's name changed to Rikke Brinkø Berg from Rikke Brinkø in August 2017):

Brinkø, Rikke: Realising the potential of shared space in facilities management. PhD thesis. DTU Management Engineering. April 2017.

The typology has been developed during the PhD project and a first version is featured in a scientific article:

Brinkø, Rikke, Susanne Balslev Nielsen and Juriaan van Meel: Access over ownership: A typology of shared space. *Facilities*, Vol. 33, No. 11/12, 2015, pp. 736–751.

The final version is also presented in a scientific article:

Brinkø, Rikke and Susanne Balslev Nielsen: The characteristics to consider in municipal shared spaces. *Journal of Facilities Management*, Vol. 15, No. 4, September 2017, pp. 335–351.

Use of the typology and guide – assessed by Rikke Brinkø Berg and Per Anker Jensen

As can be seen from the table that follows, the typology for shared use of space and the guide to creating shared space in municipalities are especially well suited for use in three of the five processes presented in Part A: strategy development, space planning and building project. The typology can be used as a concept-and-analysis tool, for example, together with typology of office workplaces see Chapter B.1, and for creativity and facilities, see Chapter B.2. The guide to creating shared space can be seen as a planning tool, for example, together with value-based space optimisation (see Chapter B.4). The guide has been developed with a specific focus on municipalities but can also be used in the creation of shared spaces in other types of organisations, with minor adaptations. The cases can be used as inspiration for a deeper understanding of the concept of shared space.

Process	Phase							
Strategy development	A	B	C	D	E	F		
Organisational design	A	B	C	D	E	F		
Space planning	A	B	C	D	E	F	G	H
Building project	A	B	C	D	E	F	G	H
Process optimisation	A	B	C	D	E	F		

For strategy development, the typology can be used by senior executives and employees for analyses of the current situation, strategic strategy definition and development of strategic planning (phases B–D), while the guide can also be used in the development of strategic plans as well as during implementation and reassessment (phase D-F).

For space planning, the typology and the guide can be used by strategic and tactical staff responsible for space management internally in the organisation through all phases from mapping out the status of the company's current supply and demand of space of different categories and for re-evaluation of the use of space (phases A–H). In addition, the typology can be used as a basis for the dialogue with FM management and departments affected by the decision (phase D) as well as with the affected departments and external partners during the planning, implementation and evaluation of redeployment (phases E–G).

For building projects, the typology and the guide can be used by managers and employees in building client and FM functions at the strategic and tactical levels during pre-project phases with the initiation of building projects and strategic programming (phases A–B) as well as in the early project phases with building programming and design (phases C–D). In addition, the tools can be used as a basis in the final post-project phase with evaluation (phase H).

B.II

Sustainability from goal to action

B.6 Strategies for sustainable FM

Susanne Balslev Nielsen

Introduction

FM professionals can play a significant role in reducing the world's major resource challenges and create sustainable development, if sustainability (environmentally, socially and economically) is integrated as a quality goal in FM at strategic, tactical and operational levels. The purpose of this chapter is to provide facility managers with inspiration for formulating sustainability strategies. The chapter points to three main directions in sustainability strategies and provides an example of how DTU Campus Service started its sustainability work.

Companies must necessarily develop their own sustainability strategy to reflect the company's values and ambitions. With strategy, I think, as the management guru Mintzberg, about perspective. It is not only about FM organisations having reached a certain position from which they can claim to be sustainable, but more about the perspective they have for their continuous development work. In my research, I have analysed examples of the FM industry's efforts to integrate sustainability, and in spite of similarities, the diversity is striking. The differences are due to the fact that companies differ in core business and facilities, but they also differ in the way they relate to sustainability and in the role and responsibility of the FM organisation. At the operational level, there are often similarities between what you actually do, for example, energy optimisation, space optimisation, green procurement, but the strategic and tactical differences imply major differences in how much and how to invest in sustainability, and how determined the organisation is if unpredicted disadvantages occur during implementation and operation.

Four positions and three main directions

In the book *The Necessary Revolution* by Senge et al., the authors present a model that is used to describe four strategic positions that characterise how organisations relate to sustainability. The four categories arise from differences in focussing on their own internal organisation (e.g. 'mind your own business' first) or assuming sustainability requires cooperation with external parties (such as United Nation collaboration or Global Mayors Forum). The second dimension involves you

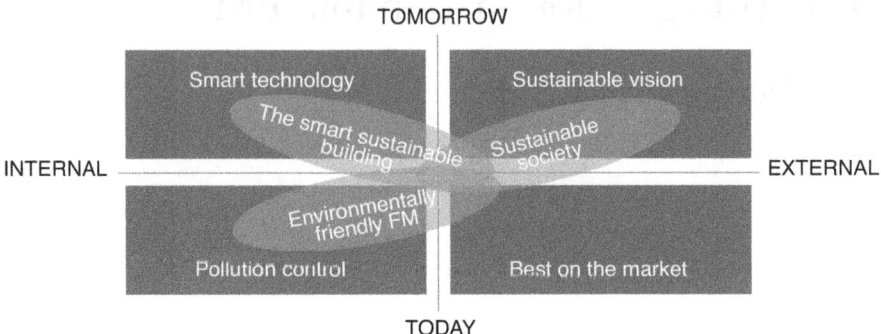

Figure B6.1 Three main directions for sustainable FM (SFM) and four strategic positions

primarily basing your decisions on the situation of today (e.g. current energy supply and prices) or basing your decisions on expectations about the situation of tomorrow, meaning the situation in the future. These four strategic positions are shown as dark rectangles in Figure B6.1. In the following, the strategic positions are discussed more closely.

Pollution control: Most organisations that have worked for a long time with environmental management, energy conservation and environmental considerations can be found in this category. Here there is a focus on reducing the environmental impact of the organisation's activities, and documentation, analysis and action plans are being used to reduce the organisation's environmental impact. These are, for example, housing associations working with green accounts that were already active with Agenda 21 activities in the 1990s and who strive to reduce environmental impact through behavioural changes and technical measures.

Smart technology: For other organisations, the focus is on the use of new technology and helping to change society by applying new technological solutions. This could be IT systems that optimise logistics, smart energy systems or other installations in buildings that reduce environmental impact. In the FM industry, there are companies that provide new technological solutions (products and services) that show a way to a future without the use of fossil fuels, etc.

Best on the market: These organisations want to be leaders in their product, and at a time when sustainability is high on the community agenda, it includes competing for sustainability. High marketing evidence shows that CO_2-neutral, emission-free and CSR activities are used in marketing, with the expectation of benefitting both society and organisation. They see business opportunities by being first and better than their closest competitor.

Sustainable Vision: The last type of organisation is one that engages in the transformation of society based on a vision of a sustainable society and a future that depends on making partnerships with peers and experts. Here the challenge is seen as social and structural, and there is a focus on reconsidering ways of living, building and working.

Three main directions for the realisation of sustainable FM

Sustainable FM (SFM) is still something that exists more in theory than in practice. But there are examples of sustainability strategies, action plans and projects, and I have identified three main directions regarding the realisation of sustainable FM. These are shown by the oval areas in Figure B6.1.

Environmentally friendly FM: FM organisations who know their environmental impact (e.g. environmentally certified) have focus on minimising their environmental impact and, if possible, contribute to biodiversity and material reuse.

The smart sustainable building: These are most often seen in new construction and larger building renovations. The goal is to create buildings that are good for the users of the buildings and the biodiversity on the premises. In the strategic goals, they focus on environment and economy, including internal social values, such as a good indoor climate (light, air, noise, particles, etc.) and universal design.

Pilot projects for a sustainable society of the future: Some FM organisations have, in addition to managing their basic and internal functions, the perspective that they will take co-responsibility for developing new and sustainable solutions. The strategic goals also emphasise external social values, such as nature conservation and working conditions at local and global level. Here time and resources are spent on knowledge sharing and development projects that can reconsider facilities and FM.

Using innovation terminology, the aforementioned three strategic directions are, respectively, a gradual, a radical and a transition's strategy within FM. The study of examples of sustainability in FM has led to models that can inspire others for their own efforts. But it is one thing to describe how I see the concept of sustainability integrated into FM today and another to comment to what in the short and long term leads to the most sustainability. Immediately, environmentally friendly FM is a minimum effort, but its effect should not be underestimated, as FM can do a lot to reduce the environmental impact, especially from the existing building. The smart sustainable building is ideally more holistic and includes sustainability in a broad sense, but the restriction may be the creation of sustainable islands, which in themselves are positive, but which can also close off for contributing to sustainable urban development in the local community. Immediately, the perspective of social change is the biggest and most ambitious perspective. However, the effects may be more long-term and subject to greater risk and uncertainty.

What is your and your FM organisation's sphere of influence now and in the future?

Another significant framework for sustainable FM is linked to the perception of FM and the role FM organisations are allowed to have in terms of supporting the organisation's core business. A group of researchers (Pathirage et al.) has described four generations of FM over the past 30–40 years, which vary in terms of perception of purposes and main tasks for in-house FM organisations. The generation that is practised in an individual company is important for how the sustainability challenge is interpreted and thus how a facility manager can argue for investments to improve the company's sustainability profile.

- *1st Generation:* FM is seen as an expense to be driven for minimum expenses rather than optimal value.
- *2nd Generation:* FM is seen as an integrated continuing process in relation to the company's development.
- *3rd Generation:* FM is seen as resource management focussing on supply chain management in relation to FM functions.
- *4th Generation:* FM is seen as strategic management to ensure coherence between organisational structure, work processes and the physical framework and the organisation's strategic goals.

If you are in a company where your leaders focus solely on financial management (1st generation FM), the specific actions can only be realised if they meet the economic goals of profitability, while social and environmental benefits do not count as independent arguments. However, if it is a question of how FM in all its diversity can contribute to the business's work, the well-being of the employees and society-level development (4th generation FM), there are completely different and broader scope for achieving sustainable FM.

What approach do you have for sustainability in FM in the given context of your organisation? The strategic positions described can help with an analysis of your organisation and possible SFM approaches. The strongest approach will probably occur if you combine elements from the various approaches and take advantage of the opportunities that arise in the form of ongoing operational improvements and larger projects.

Case of sustainability at DTU Campus Service

This chapter is about FM organisations having various strategies for sustainable FM. The following is an example of actions taken by an FM organisation. The example illustrates that work with sustainability in FM should be seen as a continuing development journey rather than a completed journey towards a determined destination.

Engineers are trained to create improvements in society, and at a university like the Technical University of Denmark (DTU), sustainability is high on the political agenda in terms of education, research, innovation and research-based consulting. DTU Campus Service (CAS), the university's FM department, is a major internal partner in achieving sustainability at the university. Therefore, CAS has begun a development process that will lead to the systematic integration of sustainability considerations with other functional, economic and aesthetic considerations in a new and overall quality objective. As part of a research project on sustainability in FM, I have had the opportunity to follow this start-up phase closely. Since 2011, CAS's sustainability work has continued; therefore, the image I am drawing here from CAS is only a snapshot from an earlier stage.

Sustainable FM is about long-term holistic thinking instead of partial solutions and short-term thinking. CAS has very good prerequisites for long-term holistic thinking because it is both a building owner and responsible for day-to-day operations. DTU is one of the only universities in Denmark which owns its buildings. Sustainability in FM at DTU is relatively new. The university campus, initially, was completed in 1974, and until 2001, the strategic focus was on building operations, and there was a situation characterised by ample capacity, especially in the beginning. In 2001, the building ownership was changed from the state to DTU, and this marked the beginning of more than a decade of investments transforming the Lyngby Campus with building renovations and new buildings. Around 2009, CAS was reorganised, making it more and more an FM organisation, where it used to have building operations as its only focus. Energy efficiency has been an issue ever since the energy crisis in the 1970s, but the dedication to sustainability in CAS started in 2009, and in 2011, DTU's first sustainability policy was approved.

To describe how SFM is realised, a strategy model was developed by the aforementioned Mintzberg et al. in the book *Strategy Safari: A Guided Tour through the Wilds of Strategic Management*. The strength of this theory is that it gives words to describe why good intentions are not always realised and that there can be external developments that align with the organisation's strategic intentions but without being the result of an initiated strategy process (see Figure B6.2).

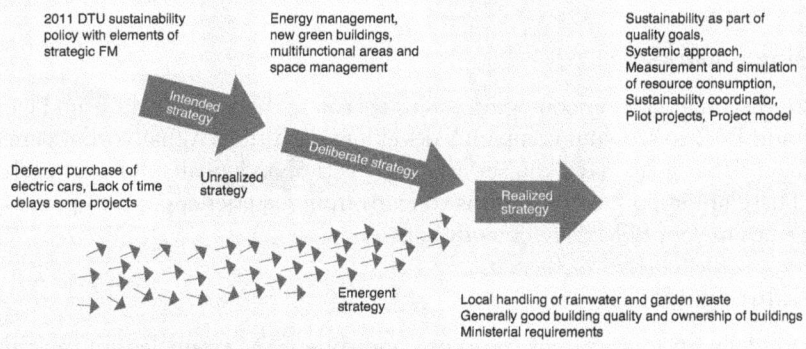

Figure B6.2 The journey to realised sustainability in FM at DTU

The intended strategy: In this case, this is DTU's sustainability policy, which is a reference to the annual development and action plans in CAS. The document states intentions: 'DTU's sustainability policy supports the university's vision of being a globally recognised, elite of technology'. The concept of sustainability is defined as a holistic term, which includes an environmental, social and economic dimension. The policy dictates a system-based approach, meaning that, for example, building energy and process energy, individual buildings and energy systems, resource consumption and recycling are to be seen as a whole.

The deliberate strategy: This strategy represents the dedicated effort that an organisation is working on. In this case, their main efforts were to get a detailed overview of energy consumption, to ensure sustainability in building briefs for the new buildings and to minimise the need for new land by intensifying the use of existing areas.

The unrealised strategy: With Mintzberg's terminology, this is the part of the planned strategy which is not realised yet. As the planned strategy for DTU was relatively general, it was not clear what would not be realised. For example, for a period of time, it was decided that they would not invest in electric cars because the investment at that time seemed too uncertain. In addition, employees' time pressure meant that, for example, there were energy management projects that were not realised as soon as expected.

The emergent strategy: This strategy is a result of external requirements and initiatives that are currently seen as part of the University's sustainability profile, but have been initiated by somebody else than DTU. For example, Lyngby Campus was built with infiltration trenches for storm water, something that adds scores in sustainability certification. For many years, they have made green accounts and some university departments have chosen to work with behavioural campaigns themselves. In the meantime, technological development has made many components more energy efficient. This also contributes to the realised sustainability at DTU.

Closing remarks

I close this chapter by encouraging you to try to use the frameworks from Figures B6.1 and B6.2 to tell your company's development history. What is your general strategy? What goals have you set and what action and results did you reach so far? These are important reflections to learn from experiences and to plan your next steps in your SFM development journey.

Literature guide

The model with four strategic positions and three SFM strategies are treated in more depth in the following book chapter:

Nielsen, Susanne Balslev: Claims of sustainable FM: Exploring current practices. Chapter 4.2 in Per Anker Jensen and Susanne Balslev Nielsen (eds.): *Facilities*

Management Research in the Nordic Countries: Past, Present and Future. Centre for Facilities Management – Realdania Research, DTU Management Engineering, and Polyteknisk Forlag, January 2012b.

The case about DTU Campus Service is treated in more depth in the following conference paper:

Møller, Jacob Steen, Susanne Balslev Nielsen and Keith Alexander: Sustainability in higher education facilities management: A pilot study of at the technical university of Denmark. Proceedings of the 11th EuroFM Research Symposium: 24–25 May in Copenhagen, Denmark. Centre for Facilities Management – Realdania Research, DTU Management Engineering, and Polyteknisk Forlag, May 2012.

Use of the tool – assessed by Per Anker Jensen

As shown in the table that follows, strategies for sustainable FM are particularly useful for three of the five processes presented in Part A: strategy development, space planning and building project. The tool can be used as a terminology and analysis tool. The case can be used as inspiration to work with strategies for sustainable FM.

Process	Phase							
Strategy development	A	B	C	D	E	F		
Organisational design	A	B	C	D	E	F		
Space planning	A	B	C	D	E	F	G	H
Building project	A	B	C	D	E	F	G	H
Process optimisation	A	B	C	D	E	F		

For strategy development, the tool can be used by managers and staff on the strategic level for analyses of the current situation, defining strategy goals and development of strategy plans (phases B+C+D) and with re-evaluation (phase F).

For space planning, the tool can be used by managers and staff on the strategic level responsible for space management for evaluating the sustainability aspects in the internal work through the initial phases from the mapping of status for the company's current demand and supply of space of different categories to decision (phases A–D). In addition, the tool can be used as a basis for the dialogue between FM management and the affected departments about decisions (phase D).

For building project, the tool can be used by managers and staff in client and FM functions at the strategic and tactical levels to evaluate sustainability aspects during pre-project phases with the initiation of building projects and strategic briefings (phases A+B) and in the early project phase with building briefings (phase D).

B.7 Capacity building in FM organisations

Kirsten Ramskov Galamba and
Susanne Balslev Nielsen

Introduction

To set the context of this chapter, we will introduce a definition of the term profession. According to a Danish open encyclopaedia (Gyldendal), the word profession is used for 'a profession whose practitioners have a background in a specific formal education that gives them academic authority and status'. A profession in this sense is characterised by professional standards and standards for well-executed work, and in some cases also a professional ethics. Professionals belonging to a specific profession are thus guided by and refer to the norms and values of the profession rather than the system they are a part of. Autonomy is, therefore, a word often associated with professions.

However, the word professionalisation is widely used in today's society, referring to the way organisations and thus employees are being urged to work more efficiently and 'professionally'. According to critical researchers in public, neoliberal governance, professionalisation of organisations is happening at the expense of autonomy of professions and for the gain of the systems effectivity. The old professions are increasingly governed by market principles, such as the decentralisation of budgets, performance indicators and benchmarking, and decisions are made with reference to efficiency as the main criteria. At the same time, more and more professional groups are included in the name of professionalisation, which within Facilities Management could be, for example, 'professional cleaning' – a service area performed by unskilled workers and thus not belonging to a profession.

In professionalisation processes, the individual employee is expected to develop personal behaviour in accordance with the organisation's goals and values. When these behaviours are efficiency, flexibility and customer service, the employee must sometimes compromise professional values to be able to fulfil management's goals. From a managerial point of view, professionalisation can thus be seen as a way of managing trained professionals through the development of new norms and values that hold the efficiency of the organisation over that of the autonomy of the profession. Furthermore, new employee groups are 'professionalised' to create new self-governing functions focussing on quality and customer needs at the centre.

Facilities Management as a profession

In Denmark, development of Facilities Management (FM) as a professional field is on the agenda of the Danish Facilities Management Network (DFM). The network focusses on 'developing the subject area, spreading and exchanging knowledge about Facilities Management, promoting interaction between practice, education and research and being a link to international development in the field'. In the following section, we will relate the theoretical framing noted earlier with DFM's understanding of FM.

With reference to the aforementioned theoretical investigation, FM can be seen as a 'professionalised unit' composed of unskilled professionals and trained professionals with roots in their respective academic standards and values. This is confirmed by the content at the Danish Facilities Management network's website, where FM is described as an umbrella across a number of disciplines that subscribe to the following values: 'the customer at the centre' and 'optimisation of processes'.

These values are also keywords for neo-liberal professionalisation:

> Facilities Management supports the framework for the employees' primary tasks, thus providing the employees with the best possible conditions for carrying out their daily work. Facilities Management visualises which processes can be optimized and how to do it. Concepts such as flexibility, optimisation in services and systematics are included, all with reference to the company's bottom line.

FM does furthermore, according to DFM, add value:

> Experience shows that there is great gain from developing Facilities Management actively. It is possible to map strengths and weaknesses of workflows and often the quality of performed tasks can be improved. There is thus added value for the company when professional Facilities Management is developed to manage daily workflows.

But what is the consequence of always subscribing to values such as 'the customer at the centre', 'efficiency', 'quality in service delivery' and 'added value'? Does quality mean that the customer is satisfied or that the work is carried out properly according to the profession's (old) standards? Does efficiency mean that it is getting cheaper – or of a higher professional standard? Does the facilities manager work with a short-term or long-term perspective? These are just some of the issues that relate to FM as a collective category for a number of old professions and newly professional services.

What is the impact of the professionalisation processes for the FM industry, business and public institutions – and for society as a whole?

From a system optimisation perspective, there are thus quite unambiguous advantages in working towards the professionalisation of FM through the development of skills and the development of personal (professional) behaviour that constantly focusses on the customer and strives for system efficiency.

From a societal point of view, however, 'the customer at the centre' and 'system efficiency' are not necessarily values that bring society forward in a sustainable direction. In the professionalisation process, the skills and values of the old professions are challenged, and employees must constantly balance managerial demands of efficiency and professional assessments based on the values, norms and ethics of the profession with the risk of ending up with short-term solutions that are less sustainable in a societal perspective in order to meet quantitative performance demands and rigid budgets.

The ever-increasing demands for efficiency and consequent focus on lean management, time studies, etc., draws on the thinking from scientific management/Fordism. Facilities managers working at tactical or strategic levels must furthermore navigate values such as flexibility and service mind-set, which is reflected in 'post-Fordistic' trends with a demand for increased professionalisation as a managerial response. The monotonous, fragmented work tasks, as well as the flexible, value-driven work, have the goal of efficiency, and in both cases, the professional scope for the FM employee is often restricted to tasks that can be planned and evaluated through market-like mechanisms.

Not all municipalities have subscribed to FM rhetoric. Adaptation to steering technologies within the new public management governance paradigm, with the introduction of market-driven steering mechanisms as, for example, performance contracts and KPIs, means that many employees in municipal property departments will recognise the critique elaborated upon so far.

Case

As part of Kirsten Ramskov Galamba's PhD project, an action research project was conducted in cooperation with the Department of Property, Road and Park in the Danish municipality of Albertslund. The aim was to explore the concept of Sustainable Facilities Management (SFM), focussing on how to build capacity in the organisation to work meaningfully with sustainability from an FM position. Along the way, the challenges employees are meeting when working with sustainability became clear. At the same time, it provided insight as to the understanding of sustainability that Facilities Management employees subscribe to.

Regarding the possibility of change towards a more sustainable practice, it proved to be extremely difficult for the FM employee to take decisions and act beyond the daily demands from users. This means that changing practice in a sustainable direction is difficult because strategic thinking and development is something that one must find the time to do after the demands for urgent tasks have been met – and will they ever be?

Model for FM competencies for sustainable development

Based on knowledge from the action research process in 2008–2011, it has been possible in general terms to describe what is needed to build capacity to address an understanding of sustainability beyond resource efficiency and energy savings.

The action research process in Albertslund Municipality's FM organisation resulted in a more nuanced understanding of knowledge needs and various ways to strengthen control of their practice. Although a common code of conduct was not explicitly formulated in the Albertslund Municipality process, it became clear to everyone that it takes a certain mind-set to address sustainability issues in daily practice.

Employees referred to the fact that working holistically and strategically with SFM requires knowledge and an understanding of a sustainability that extends beyond 'energy efficiency'. In addition, an understanding of implications and strengths in various governance and associated technologies strengthens the ability to navigate the system. As an example, a project with a new kindergarten was mentioned, which should be built in accordance with the German Passive House Standard. The project team needed knowledge and skills related to the project's phases and the following operation and maintenance of the building – as well as understanding the passive house standard and its technical implications. In addition, team members had to understand the municipal decision-making processes, criteria for priorities, framework for funding, political ambitions, etc.

It seems obvious that facility managers should have control over their own practices at all times. However, in the Albertslund Municipality, this was not the case at the beginning of the action research process. Employees were unable to prepare long-term plans, and daily work was determined by the most urgent tasks. The consequence was slow work procedures, uncoordinated initiatives, sub-optimisation and frustration among employees. When this situation was discussed from a working perspective and the current practice was analysed from an outlook based on the utopian-future horizon for a sustainable working life, employees developed a concept of 'free arenas'. Free arenas are space (time and place) where innovation can occur without being guided by specific goals or KPIs – a break from the daily work for common reflection and development. In addition, the action research process resulted in knowledge about different ways to increase control of one's own practices through explicit strategies and planning, collaboration and knowledge sharing, as well as education and training.

Since the action research project, the capacity-building model has been further developed as presented in Figure B7.1. The model, which can be used as a process tool, outlines important elements to discuss in order to build strategic capacity in the FM organisation.

A clear strategic goal is a prerequisite for long-term planning, enabling the formulation of clear criteria for well-executed work (quantitative KPIs/qualitative values). In addition to goals for the future development of the organisation, individual facility managers need explicit knowledge in different areas, such as know-how related to facilities management disciplines and a

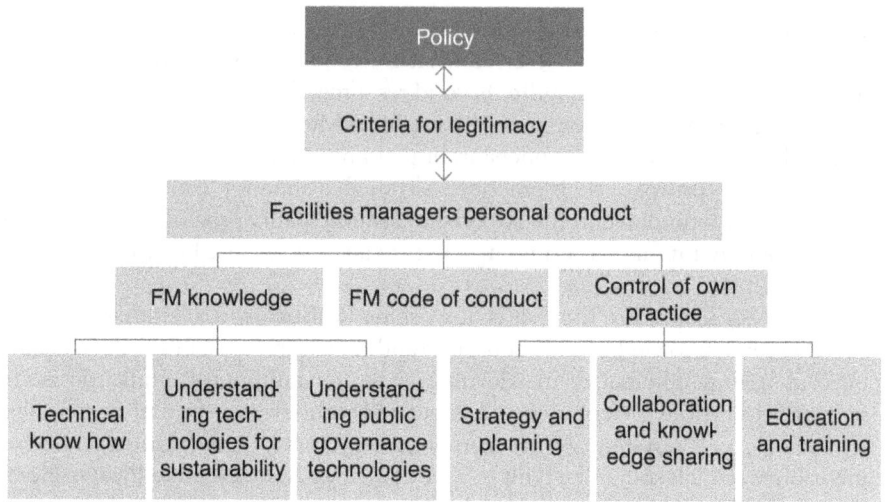

Figure B7.1 Capacity-building model for FM organisations

(critical) understanding of sustainability technologies and of public governance technologies.

Furthermore, in order to actually experience control over his or her own practice, the Facilities Management employee must have knowledge of current strategy and be able to plan accordingly. There must be a culture supporting knowledge sharing and cooperation across professions, and there should be a strong focus on education and training of employees to ensure up-to-date knowledge and skills in the organisation.

Finally, a common code of conduct for the FM organisation can support a culture focussing on, for example, service, quality, coordination, collaboration – or what is considered important to achieve the organisation's strategic goals.

Literature guide

The chapter is mainly based on the following scientific article:

Galamba, Kirsten Ramskov and Susanne Balslev Nielsen: Towards sustainable public facilities: Collectively building capabilities. *Facilities*, Vol. 34, No. 3/4, 2016, pp. 177–195.

A comprehensive presentation of the PhD project can be found in the following PhD thesis:

Galamba, Kirsten Ramskov: Public facilities management and action research for sustainability. PhD thesis 6.2012. DTU Management Engineering. May 2012.

Kirsten's CFM publications also include the following:

Galamba, Kirsten Ramskov: A critical reenvironmental management system as a tool for sustainability. Chapter 4.3 in Per Anker Jensen and Susanne Balslev Nielsen (eds.): *Facilities Management Research in the Nordic Countries: Past, Present and Future*. Centre for Facilities Management – Realdania Research, DTU Management Engineering, and Polyteknisk Forlag, January 2012.

Nielsen, Susanne Balslev and Kirsten Ramskov Galamba: Facilities management: When sustainable development is core business. *EFMC2010 Conference and Research Symposium, Madrid, 1–2 June 2010.*

Nielsen, Susanne Balslev, Anna-Liisa Sarasoja and Kirsten Ramskov Galamba: Sustainability in facilities management: An overview of current research. *Facilities*, Vol. 34, No. 9/10, 2016, pp. 535–563.

Use of the model – assessed by Per Anker Jensen

As shown in the table that follows, the model for capacity building in FM organisations is particularly useful in one of the five processes presented in Part A: organisational design. The model can be used as a dialogue and analysis tool. The case can be used as inspiration to work with FM competences for sustainable development. Even though the model is developed based on action research in a municipality and specifically focussed on public organisations, it can also be used in other types of organisations.

Process	Phase							
Strategy development	A	B	C	D	E	F		
Organisational design	A	B	C	D	E	F		
Space planning	A	B	C	D	E	F	G	H
Building project	A	B	C	D	E	F	G	H
Process optimisation	A	B	C	D	E	F		

For organisational design, the model can be used by managers and staff on all levels from analyses of purpose through all phases in between to follow up and adjusting (phases A–F). The results of capacity building in FM organisations is that the staff will be better qualified to take an active part in strategy development – thus the model can indirectly contribute to process strategy development.

B.8 Climate change and buildings

Rimante A. Cox

Introduction

Despite international attempts to prevent climate change by reducing green-house gas emissions, it is now clear that a certain level of global warming is inevitable. Therefore, there is now considerable interest in investigating, how different countries can adapt to climate changes. For example, there have been some studies of the consequences for our health and well-being, as well as the effect of climate change on agricultural production and ecosystems. In this project, we have dealt with the effect of climate change on buildings, and we offer a method of how FM could respond to it.

One might expect that a general increase in the global temperature of between 2 and 4 degrees will not seriously affect our buildings. However, climate models for the future also predict an increase in extreme weather events, such as heatwaves, prolonged droughts and intense rainfall, which often cause floods, as well as storms and hurricanes. These extreme weather conditions can cause significant damage to buildings and infrastructure. An example of this has already been experienced during the extreme rainfall in July 2011 in the Copenhagen area, where roads and residential areas were flooded.

To understand how buildings are affected by climate change, there is a need to develop local climate models that can predict temperature fluctuations (not only average annual or monthly temperatures but also daily or hourly), the likelihood of extreme rainfalls and their intensity, how wind intensity will rise in the future and how it will affect flooding in the premise's areas. The Danish Meteorological Institute (DMI) has advanced climate models to predict extreme weather conditions in Denmark. However, the data required by engineers – for example, for calculation of future energy demand – requires further processing. To produce such data typically requires down scaling global circulation models to the regional level – e.g. using regional climate models – followed by detailed analyses to assess the quality of the projections. This work requires expert knowledge of climatology, and it is typically conducted at dedicated research centres and national metrological institutes. If such data is not available, an estimated future heating and cooling demand can be calculated by using a coarse annual estimate of temperature difference as it was demonstrated in my PhD thesis.

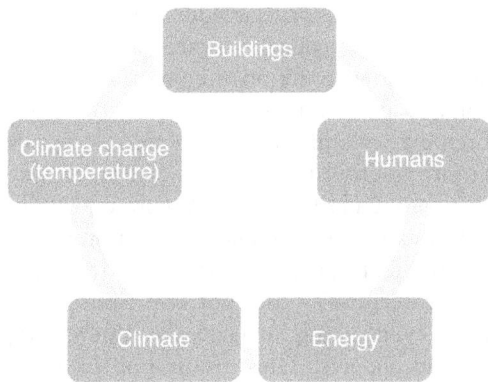

Figure B8.1 Relationship between buildings and climate change

The relationship between climate change and buildings is illustrated in Figure B8.1. Buildings protect people from the surrounding environment, from weather and changing climate. Buildings are also a product of human activity: the way we choose to design, operate and maintain our buildings has an effect on the surrounding environment. Therefore, one question is should we adapt our buildings to climate change, or should we also mitigate, which means to reduce the impact of buildings on climate change?

In the project, we have distinguished between adaptation and mitigation strategies. Adaptation strategies deal with symptoms of climate change, such as increased temperature, and mitigation strategies, on the other hand, are also concerned with the underlying cause. For example, when the average temperature rises, additional adaptation to the cooling system will be added according to the adaptation strategy. In that case, buildings will use more energy for cooling. The increased energy demand, if provided by fossil fuels, exacerbates global warming, increasing average temperature further. This, in turn, leads to further increases in energy consumption in order to cool the buildings. And a perpetual loop is created, where the immediate solution subsequently makes the problem worse. To avoid this unsustainable feedback loop, it is important to not only treat the immediate symptoms but also the underlying causes of climate change.

In this context, a mitigation strategy is a strategy that not only deals with an immediate problem but also considers how different solutions can affect the future environment. Thus a building cooling mitigation strategy recommends installing passive cooling or ensuring that the energy consumption of air conditioners is provided by sustainable, CO_2-neutral sources.

Another example of the difference between adaptation and mitigation strategies can be illustrated by different solutions in connection to flooding problems. By treating walls with waterproof materials or establishing a degree of drainage, you can solve the immediate flooding problem. However, these solutions may lead water to the surrounding buildings or charge local drainage systems. With a mitigation strategy, one can work to reduce rainwater drainage by, for example, using a green roof or gathering rainwater for on-site use or increasing soil drainage.

The environmental impact of climate change is also an important factor in the operation and maintenance of buildings as they make a significant contribution to the company's environmental impact. There is great potential in reducing environmental impact through FM. While FM represents only a small fraction of an enterprise's budget, it can have a significant effect on the company's environmental impact. For example, FM expenses according to Finnish researchers represent 4%–6% of an enterprise's costs but contribute 53%–82% to overall environmental impact, even taking into account different types of companies, their sizes, locations and budgets. Thus, the use of sustainable FM will be an opportunity to reduce a company's environmental impact and support the company's sustainable image.

Today, mitigation strategies can sometimes be more expensive than adaptation strategies. But we expect this to change in the near future, when different governments around the world introduce new building regulations and taxes on any greenhouse gas emissions. In order to investigate how mitigation strategies can be used in practice, we have worked with Gentofte Municipality, where we have applied mitigation strategies for the renovation of historic buildings.

Managing the impact of climate change on buildings

A result of the project was to develop a decision support tool for facilities managers to quantify the total cost of assessing measures to address the impact of climate change on buildings. Theoretically, the tool is based on coupling and quantifying sustainability and resilience. By resilience we mean that we understand how well a building or system continues to function during and after an event. There are many definitions and aspects of resilience, but our focus is on planning, preparing and protecting measures to increase the level of resilience.

To couple resilience and sustainability, we propose determining the expected cost (risk) to a company for providing a function of a building in a particular manner, where the cost considers functional (resilience) as well as environmental, economic and, possibly, social dimensions (sustainability). This is very different from looking at resilience alone. For example, if we are looking at the cooling of a building, resilience only considers under what conditions the cooling systems will fail and the expected costs associated with the loss of the function. If there is no loss of function, the resilience is perfect, and there are no expected costs. In contrast, the combination of resilience and sustainability considers not only the costs associated with loss of function but also costs related to environmental and economic sustainability. Thus the expected costs associated with resilience and sustainability of say cooling may be high, even when there is no loss of functionality if providing this function has environmental, economic or social repercussions.

Our method, called Coupling and Quantifying Resilience and Sustainability (CQRS), consists of the following seven steps:

1 Determine the resilience of the building to the disturbance, e.g. at what temperature the building's function will be compromised.
2 Determine the costs associated with both the loss of the building's functionality and the building's current sustainability.
3 Determine the corresponding probabilities associated with each cost.
4 Determine the expected costs associated with the current resilience and current sustainability of the building using risk analysis.
5 Determine the capital and operational costs of each remedial solution.
6 Determine the expected cost associated with the proposed resilience and proposed sustainability of the building using risk analysis for each remedial solution.
7 Select (or not) a solution based on cost-benefit analysis.

An illustrative example

We illustrate the CQRS method in a hypothetical case. We look at an architectural company called Green, Greener & Greenest (GG & G), which specialises in designing and refurbishing environmentally sustainable buildings. GG & G's office building is located in an urban environment and consists of a three-storey building originally built in the 1920s. The building provides office space for 30 employees. It is centrally heated and has a small central air conditioning unit. In the previous year, GG & G experienced a number of days where the building was uncomfortably hot due to an unusually warm summer period. During that time, the air conditioning system was not capable of cooling the building sufficiently. As a result, the company lost 100 working hours. The company, therefore, decided to investigate their resilience and sustainability to heatwaves.

Using the CQRS method, GG & G calculated that the existing situation (do nothing) would cost the company $130,000 over a five-year period, where $112,000 was due to the loss of function and $18,000 was the operational cost of the existing system. Using the same method, the company investigated following two alternative solutions:

1 Upgrading air conditioning systems
2 As Solution 1, supplemented with solar power plants for electricity production

If only the resilience to the heatwave was considered, in Solution 1, the total cost was slightly lower than in the existing system and was $121,600, of which the investment of the upgrading of the system was $100,000, while operating expenses increased modestly and cost of loss of function were eliminated.

In Solution 2, the total cost was $130,000, as in the existing situation, as the investment in the solar system amounted to $30,000, while the operating expenses

Table B8.1 Expected cost for the company over a five-year period when both resilience and sustainability are considered

CQRS method	Existing ($)	Solution 1 ($)	Solution 2 ($)
Capital cost of a/c	0	100,000	100,000
Capital cost of PV	0	0	30,000
Expected annual cost due to loss of reputation	131.000	185,000	0
Expected annual cost of carbon tax	1.260	1,512	0
Loss of function	112.000	0	0
Operational cost	18.00	21,600	0
Total	262,260	308,000	130,000

declined, as the electricity produced could cover the supply to the air conditioning system.

However, when costs associated with sustainability were recognised, the result was significantly different, as shown in Table B8.1. In the existing system, the total expense now became $262,500 over a five-year period – i.e. approximately double. The main difference was an expected cost of $131,000 due to the loss of reputation as the company chose a conventional air conditioning system for a company that specialises in sustainable buildings. In addition, a lower cost of $1,260 was included in CO_2 taxes. With Solution 1, both of these costs increased, and the result was a total cost of $308,000, i.e. an increase of 18% relative to the existing situation.

In Solution 2, the result was $130,000 in the calculation for resilience alone, as there were no expenses associated with the sustainability (no loss of reputation or CO_2 taxes). This solution thus had a total cost of almost half in relation to the existing situation, taking into account both resilience and sustainability.

Literature guide

The chapter is based on the following scientific article:

Cox, Rimante A., Susanne Balslev Nielsen and Carsten Rode: Coupling and quantifying resilience and sustainability in facilities management. *Journal of Facilities Management*, Vol. 13, No. 4, 2015.

An overall presentation of the PhD project can be found in the PhD thesis:

Cox, Rimante A.: Climate change and its impact on the operation and maintenance of buildings. PhD Thesis. DTU Civil Engineering, Report No. 312. 2015.

Other publications by Rimante include a scientific article and a conference paper:

Cox, Rimante A., Martin Drews, Susanne Balslev Nielsen and Carsten Rode: Simple future weather files for estimating heating and cooling demand. *Building and Environment*, Vol. 83, January 2015, pp. 104–114.

Cox, Rimante A., Susanne Balslev Nielsen and Carsten Rode: Sustainable resilience in property maintenance: Encountering changing weather conditions. *Proceedings of CIB FM Conference, Technical University of Denmark, 21–23 May 2014.*

Application of the tool – assessed by Per Anker Jensen

As shown in the table that follows, the CQRS method incorporating climate change impact on buildings is particularly suitable for use in two of the five processes presented in Part A: strategy development, space planning and building project. The method can be used as an analysis and decision support tool. The case can be used as an inspiration to work with climate change impacts on buildings. Although the method has been developed in collaboration with a municipality, it is not specifically aimed at municipal buildings.

Process	Phase							
Strategy development	A	B	C	D	E	F		
Organisational design	A	B	C	D	E	F		
Space planning	A	B	C	D	E	F	G	H
Building project	A	B	C	D	E	F	G	H
Process optimisation	A	B	C	D	E	F		

In strategy development, the methodology can be used by managers and employees at the strategic level internally in FM organisations and by their consultants to define strategy goals and develop strategy plans in relation to resilience and sustainability (phases C–D).

For building projects, the methodology can be used by managers and employees at the strategic and tactical levels internally in building client and FM organisations and their consultants in pre-project phases with the initiation of building projects and strategic briefings to define resilience and sustainability objectives (phases A+B) as well as in the early project phases with building briefing and design to define specific requirements, as well as analyse and prioritise solutions (phases C+D).

B.9 ESCO as method for learning in FM organisations

Jesper Ole Jensen

Introduction

Energy Service Companies (ESCO) is, in a Danish context, a relatively new concept; it implies that a private operator agrees to carry out (and in some case to finance) an energy optimisation of a building or other type of facility and accepts the risk for the customer having a lower energy consumption. This breaks with the hitherto way to complete energy retrofitting projects: The risk for the investment actually being paid back (through energy savings) is transferred to the private party, who in return has an influence on the solutions being chosen in order to reach the guarantee. The central idea is that the client does not buy a certain solution but a certain *energy service* – e.g. to maintain a certain quality of indoor climate (temperature and air change) for a certain cost that expectedly is lower than before if the quality is not changed drastically. The ESCO operator, therefore, typically will be taking care of the operation in the contract period in order to live up to the guarantee made to the client about delivering lower energy costs.

An ESCO supplier can be defined as

> a natural or legal person that delivers energy services and/or other energy efficiency improvement measures in a user's facility or premises and accepts some degree of financial risk in doing so. The payment for the service delivered is based (either wholly or in part) on the achievement of energy efficiency improvements and on the meeting of the other agreed performance criteria.
>
> (EU Directive on Energy End-Use Efficiency and Energy Services [European Parliament and Council of the European Union 2006])

In practice, the term 'ESCO' is differently defined across countries and covers various arrangements as Energy Performance Contracting (EPC), Energy Service Provider Companies (ESPC), Delivery Contracting (DC), 'Comfort Contracting', 'Supply Contracting', 'Energy Supply Contracting' and Contract Energy Management (CEM).

The original definition of ESCO involves third-party financing. However, this has not so far been interesting for Danish municipalities, as they are able to borrow at a much lower interest rate than private companies can offer. This is mainly due to the fact that the creditworthiness of a municipality is very large, and the lender's risk is very limited by lending to a municipality, as compared to lending to a private company. All previous ESCO contracts in Danish municipalities are thus based on the practice that the municipalities themselves finance the improvements, and the ESCO supplier is implementing the solution and guarantees a certain reduction of energy costs (and in some cases the water costs, depending on the concrete project). If the savings are not reached, the agreement entails that the ESCO provider pays the municipality the difference to the guaranteed level of savings. If more than the guaranteed saving is achieved, the municipality and the ESCO provider will share the savings according to a model defined in the contract.

The projects are typically divided into three phases (cf. Figure B9.1: analysis, including mapping; e.g. energy use in buildings), implementation of improvements and operation of the buildings. Most often, the contract includes the possibility that the municipality may terminate the cooperation after each phase, which, in practice, gives the municipality greater security when signing the contract, as a cooperation which lasts 10–12 years can be difficult to predict and therefore risky, especially since the municipalities do not know the ESCO providers in advance.

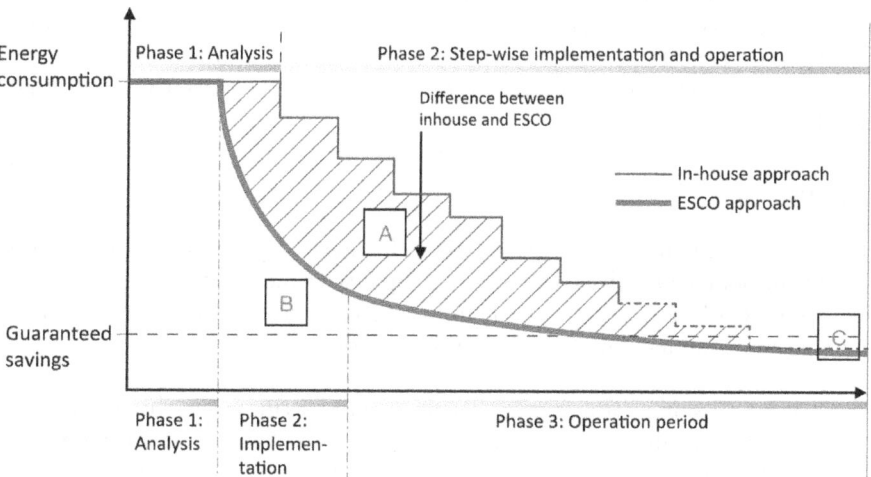

Figure B9.1 The principal difference between ESCO efforts (lower bold curved line) and an in-house energy retrofitting where the municipality itself is responsible for planning and implementing the project (upper thin stepped line), which typically takes place step by step over a number of years. The difference between the two methods consists of areas A, B and C together

Explanation on area A, B and C in Figure B9.1:

- Area A is the greater savings achieved through the ESCO model rather than the step-by-step approach.
- Area B is the difference between the ESCO provider's guarantee and actual consumption. There is often a need for adaptation of installations and systems, or local users must learn how the system works, how to achieve the most effective utilisation, etc., before the guaranteed savings are reached. This difference is paid by the ESCO provider to the client. Unlike an in-house project, the ESCO provider takes the risk that the actual savings will also meet the expected savings.
- Area C is the possible additional savings compared to the warranty, which is typically shared between the municipality and the ESCO provider.

The ESCO model typically provides a shorter analysis period and implementation period, which means that energy savings are achieved faster. The in-house model, where the municipality itself is planning and implementing the energy retrofitting, usually requires a longer analysis phase and implementation phase due to staff and budget constraints, which together provide a more elaborate process, where the energy savings are achieved step by step. At the same time, it creates uncertainty as to whether the long-term goal is achieved, unlike the ESCO model, where the final target is enrolled in the contract between the municipality and the provider. On the other hand, the ESCO model is associated with a number of transaction costs that are not present to the same extent in the in-house model. At the same time, the ESCO supplier is to be paid for its services, costs that the municipality partially avoids by carrying out the task itself. However, the municipality may need to hire additional personnel or make use of external consultants.

Why ESCO?

There are principally a number of basic advantages for ESCO rather than a more traditional approach, where the municipality would typically select specific energy saving solutions to be carried out in specific buildings and then send them to tender to be implemented by private contractors:

- *Warranty:* Issuing a warranty on specific energy savings from the ESCO supplier is an integral part of the ESCO concept. This means that the client can calculate and make sure that the money you invest in energy savings 'comes home' again, thereby making it politically attractive. It should be borne in mind, however, that the guarantee implies that there is a regular negotiation on the 'baseline' – i.e. what the starting point for energy savings should be, which other changes have been made in the building, etc., which also costs the municipality resources.

- *Capacity:* If a municipal property management decides to carry out energy improvements of the municipality's buildings, it usually takes place through a step-by-step process, as the administration can only carry out the renovation of a limited number of buildings per year and because the municipality usually only allocates funds for renovations of a limited number of buildings per year. With an ESCO project, the municipality has the opportunity to carry out renovations of a larger number of properties within a few years, as the ESCO supplier is more flexible than the municipality in order to hire more workforce temporarily, which is not possible for the municipality. ESCO contracting, therefore, typically implies that energy savings are achieved faster than if the municipality itself would implement it as an in-house project, cf. Figure B9.1.
- *Skills:* The private party may have competences related to energy savings that the municipality does not possess. The ESCO supplier often has specialised knowledge – e.g. on energy optimisation of technical installations. Therefore, it often gets fast or 'easy' savings on apparently simple regulation of the heat system, electrical installations, etc., and is able to react quickly to errors and possible changes in consumption patterns.
- *Continuation:* Maintaining focus over time is also a typical consideration for ESCO projects. An argument for ESCO is to ensure that energy conservation efforts in municipal buildings are implemented regardless of whether other important municipal priorities should arise, which is a classic danger to internal municipal projects.
- *Political visibility:* With the great focus that has been on ESCO in recent years, an ESCO project probably offers more political awareness, compared with a 'traditional' internal municipal energy saving project. ESCO can also be a way of highlighting municipal power and signalling that you participate in the new concepts. Conversely, an ESCO project can also be politically controversial if, for example, it is a conviction among politicians that the municipality should be able to carry out such tasks themselves or wish to build internal competences in this area.

Learning and innovation with ESCO in the municipal building management

A topic in the project has been the possible learning and innovation that lies in an ESCO effort. Based on theories of public innovation, innovation can be seen as part of various public administration paradigms: traditional public administration, with a stable context, where users are seen as 'clients' and management follows the adopted hierarchies. In new public management, users are perceived as 'customers' and where public governance is characterised by market logic, with a competitive context. And, finally, networked governance is where users are included as potential co-producers in the delivery of welfare benefits and where public governance takes place in networks and partnerships in a highly changing context.

FM can also be seen as a public service that can function according to different rationalities in the three public administration paradigms and involves different forms of innovation: a traditional FM approach is when innovation involves step-by-step improvements but no fundamental improvements, an ESCO-based FM approach is when innovation involves a shift in the manner in which welfare services are provided and, finally, a network-based approach is when FM extends beyond local government and takes place in collaboration with local users. This approach redefines building management and implies radical innovation.

We estimate that the ESCO concept reflects an FM concept that may be within the framework of both a new public management paradigm and a networked governance paradigm. The ESCO project in the Danish municipality of Middelfart is an example of an attempt to transfer the ESCO project's energy saving learning through network collaborators to other areas, in this case for an effort that motivates local homeowners in the municipality to implement energy savings in their own buildings. This is in line with the experience of the public innovation literature that shows that the greatest innovation is achieved through collaboration in external networking – i.e. in cooperation with actors outside the institution (in this case the municipality). Another innovative example of a municipally initiated networking on the use of the ESCO model in non-municipal buildings is the Berlin Energy Agency's call to pool private buildings for an overall ESCO offer, provide building owners with impartial guidance, implement ESCO procurement and evaluate the offers – i.e. a mediating role between different actors. The concept has a strength in overcoming a number of barriers to energy conversion in private buildings (lack of knowledge, impartial advice, uncertainty about savings, risk of supply and investment, etc.) and has contributed to achieve major energy savings in a wide range of properties in Berlin.

A central issue in relation to ESCO, based on an FM perspective, is whether the municipalities' use of ESCO impairs the municipal FM function – e.g. if the ESCO cooperation is considered a simple outsourcing – or if the ESCO cooperation in contrast can be a way to strengthen the municipal property management by making the departments more visible and linking them to strategic agendas for the municipality.

There are many indications that ESCO cooperation is being exploited actively by the municipalities in many places where the ESCO projects often have great political and local awareness, thus contributing to the visualisation of the municipality's property management or FM organisation. At the same time, a centralised FM department is often a prerequisite for being able to carry out an ESCO project, as opposed to an organisation where the individual municipal departments are building managers.

The possibilities for learning through an ESCO project are illustrated in Figure B9.2.

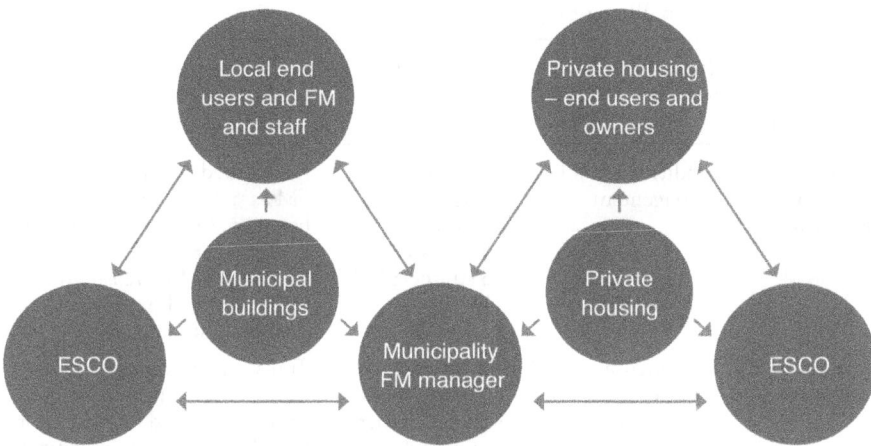

Figure B9.2 Illustration of the possible learning of the municipal FM organisation through an ESCO project, with the opportunity to further promote the ESCO approach among local homeowners

Another question is what types of innovation ESCO cooperation can help to promote: incremental, radical or transformative innovation. Is ESCO cooperation just a more effective way of implementing well-known efforts, or does it actually bring new knowledge to the organisation? Or is it fundamentally a question about the municipality – as in other ways – to get used to outsourcing, i.e. ordering and not performing? And how does it affect different parts of the organisation from management to individual operator and user?

While it may be argued that ESCO contracting has a large potential for innovation in the municipality's FM departments, the study of the Danish projects shows that there are very big differences in how innovative the municipalities experience their ESCO efforts – i.e. what learning the projects entail, both individually and organisationally. The study identified three different approaches to ESCO efforts: a basic approach, an integrated approach and a strategic approach. It is concluded that the different approaches of the municipalities to the ESCO projects reflect a high degree of flexibility in the ESCO concept, which we see as a major reason for its progress in Danish municipalities. The situation in Danish municipalities is very diverse, both in terms of building portfolio, internal resources, competences and the political context (culture, ambitions, etc.).

In Table B9.1, the earlier characteristics of different types of FM under different management paradigms with different types of innovation are compiled.

Table B9.1 Characteristics of different types of FM under different management paradigms with different types of innovation

Types of Facilities Management	Traditional FM in municipal buildings	ESCO-based FM in municipal buildings	ESCO-based FM initiatives directed towards the local housing market
Public management paradigm	Traditional public management	New public management: increasing contract steering and strategic FM	Networked governance: FM as network facilitator between, for example, private homeowners and ESCO suppliers. FM increasingly integrated with municipal sustainability initiatives
Types of innovation	Incremental: constant improvements, but no real innovation	Radical: ways of delivering service has been changed, but takes place within the usual FM framework	Transformative: experiences from ESCO contracting leads to new roles and functions for the FM unit, and to new types of organising

Literature guide

The chapter is based mainly on the final report from the project in Danish:

Jensen, Jesper Ole, Susanne Balslev Nielsen and Jesper Rohr Hansen: *ESCO i danske kommuner – En opsamling af motiver, overvejelser og foreløbige erfaringer med ESCO i kommunale bygninger*. Research Report. Danish Building Research Institute, SBi, Aalborg University. Copenhagen. 2013.

Publications from the project in English include the following:

Jensen, Jesper Ole, Jesper Rohr Hansen and Susanne Balslev Nielsen: ESCO in Danish municipalities: Experience, innovations, potential. EEDAL: *6th International Conference on Energy Efficiency in Domestic Appliances and Lighting, Copenhagen, 24–26 May 2011*.

Jensen, Jesper Ole, Jesper Rohr Hansen and Susanne Balslev Nielsen: ESCO in Danish municipalities: Basic, integrative or strategic approaches? Chapter 4.4 in Per Anker Jensen and Susanne Balslev Nielsen (eds.): *Facilities Management Research in the Nordic Countries: Past, Present and Future*. Centre for Facilities Management – Realdania Research, DTU Management Engineering, and Polyteknisk Forlag, January 2012.

Jensen, Jesper Ole, Susanne Balslev Nielsen and Jesper Rohr Hansen: Greening public buildings: ESCO-contracting in Danish municipalities. Energies, Special Issue on Energy Efficient Buildings and Green Buildings. *Energies*, Vol. 6, 2013, pp. 2407–2427.

Jensen, Jesper Ole, Susanne Balslev Nielsen and Jesper Rohr Hansen: Country report on ESCO in Denmark. Chapter in P. Langlois and S. Hansen (eds.): *World ESCO Outlook 2012*. USA: The Fairmont Press, Inc., 2012.

Jensen, Jesper Ole, Pimmie Oesten and Susanne Balslev Nielsen: ESCO as innovative facilities management in Danish municipalities. EFMC2010, Madrid, 1–2 June 2010.

Use of the method – assessed by Per Anker Jensen

As shown in the table that follows, the use of ESCO as a method for learning in municipal FM organisations is especially suitable for use in two of the five processes presented in Part A: organisational design and process optimisation. The method can be used as an analysis and decision support tool, possibly along with the method of energy-efficient FM in municipalities (see Chapter B.11). Although the method has been developed in relation to municipal FM organisations, it may also be inspiring for other organisations considering the use of ESCO.

Process	Phases							
Strategic development	A	B	C	D	E	F		
Organisational design	A	B	C	D	E	F		
Space planning	A	B	C	D	E	F	G	H
Building project	A	B	C	D	E	F	G	H
Process optimisation	A	B	C	D	E	F		

In organisational design, the methodology can be used by managers internally in FM organisations when considering organisational structure and competences in the field of energy, including in-house versus outsourcing, throughout the process from defining the purpose of organisational changes to follow up and adjustment (phases A–F), as well as in dialogue with the municipality's politicians, managers and employees on energy efficiency with or without ESCO projects.

In process optimisation, the method can be used by executives and employees at the strategic level internally in FM organisations for evaluating current performance and improvement potential, identifying, deciding and implementing changes, as well as evaluating new FM performance (phases A–D) and in dialogue with local politicians, managers and employees on the implementation of ESCO projects.

B.10 Sustainable building renovation

Per Anker Jensen

Introduction

There are many challenges and obstacles in building renovations. Many possibilities are missed in the early stages of renovation projects because of a lack of knowledge and missing economic incentives, and there is also a need for better communication between different stakeholders involved in building renovations. In order to overcome some of these issues, we developed the new value-based tool called RENO-EVALUE, which in a simple way can support dialogue about goals for renovation projects during the decision process. Furthermore, it can be used in later phases to evaluate to which degree the different criteria are fulfilled and value created during the project completion.

Value is a complex concept – particularly when we look beyond the quantifiable economic value and take qualitative aspects into consideration. A common definition is that value is the trade-off between benefits ('what you get') and sacrifices ('what you give'). From this definition follows that value has a strong subjective element. Different people will evaluate benefits and sacrifices differently. This means that we in relation to value of building renovation need to focus on the different stakeholders and their varying interests and viewpoints.

In the initial stage of the development, we did a literature study of the state-of-the art of energy renovation of buildings in Denmark as well as a needs and stakeholder analysis based on interviews with the main parties involved in energy renovation of buildings. This showed that there is a need for simple-to-use tools to support decision-making and evaluation of renovation projects. The development of RENO-EVALUE involved four case studies of renovation projects and two workshops with stakeholders. The case studies include a large social housing estate in Sorgenfri, north of Copenhagen, during the decision phase concerning a comprehensive renovation: a total renovation of a housing block completed as a pilot project in a large social housing block in Aarhus, the second-largest city in Denmark; a comprehensive renovation of a folk school in Copenhagen under completion; and an energy optimisation during and after completion aiming at raising the energy label of a privately owned office block rented out to a state organisation in Copenhagen.

Purpose

RENO-EVALUE is a tool for holistic assessment of sustainability in building renovation projects. The main purpose of RENO-EVALUE is to be a decision support tool in the early phases of renovation projects. It is a process-oriented tool, which can be used by anyone with insight into the project. RENO-EVALUE not only focuses on a final product but also covers project organisation, economy and renovation process. It can support the formulation of goals for renovation projects and enable a focus on essential aspects for the primary decision makers.

Rather than suppressing different interests, values and preferences, the tool, on the contrary, aims to make such differences visible so that they can be dealt with during the decision-making process in an open exchange of opinions, leading to real compromises and/or conscious decisions. Thus, it can be used as a communication and dialogue tool between different stakeholders in defining goals and to manage expectations in the initial phases. During the completion of projects, it can be used in an ongoing follow-up to make evaluations of the obtained results of the project compared to the defined goals and identified expectations. The tool also provides the opportunity to compare different projects and evaluate alternative proposals. Besides being a part of the development of RENO-EVALUE, the case studies are also intended to be used as illustrative cases that can provide inspiration to use the tool in connection with new renovation projects.

Target groups

RENO-EVALUE can be used by the decision makers who do not necessarily possess the adequate technical competences for evaluating energy renovation projects. The tool can be used by all stakeholders involved in the early phases of building renovation projects as long as they have some knowledge about the project.

The tool is intended for use on large-scale projects in the professional sector, not single family houses, etc. Primary users of RENO-EVALUE may be client organisations, housing associations, estate administrators, facilities managers, etc. One of the advantages of RENO-EVALUE is that it can be used as a communication tool between developers/landlords and representatives of inhabitants, tenants, employees and building users. It can also be used by project managers to manage the expectations of the different stakeholders and to show to which degree the objectives have been met. Architects, consultants and contractors might use RENO-EVALUE for illustrations and comparisons of different proposals.

Structure

RENO-EVALUE covers the four main categories: environment, users, project organisation and economy. Three of these represent the three pillars of sustainability: environment, social (users) and economy, while the last is related to the project character of renovation. Each category is divided into two parameters with a sub-division in a number of factors. The categories and the parameters are generic in relation to building types, while some of the factors are dependent on the specific building types. Factors are shown for each category and parameter in Table B10.1. This is based on the renovation of housing estates, which, for instance, is expressed in the factors for 'value'.

Table B10.1 Categories, parameters and factors in RENO-EVALUE

Category	Parameter	Factors
Environment	Resources	Energy consumption Renewable energy production Water consumption Reuse of water Reuse of building materials Amount of waste Reuse of waste
	Climate	CO_2-emissions Pollution Local discharge of water
Users	Product	Architecture and aesthetics Function and user friendliness Indoor climate and comfort Sustainability
	Process	Cooperation between participants Mutual information User involvement User consideration in implementation
Project Organisation	Developer/client	Project management skills Ability for decisions Technical competence Cooperative skills Involvement of the operating organisation Risk/responsibility/innovation
	Consultant/contractor	Project management skills Technical competences Problem solving abilities Cooperative skills Coherence in supply team Risk/responsibility/innovation
Economy	Euro/Kroner	Reasonable rent Reasonable running costs Reasonable costs in the long term
	Value	Desirable dwelling Well-functioning estate Attractive area

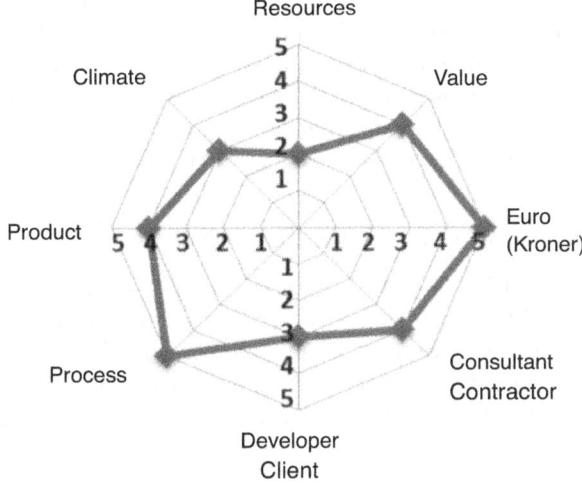

Figure B10.1 Illustration of RENO-EVALUE model

Application

Since the tool is addressing different stakeholder groups, including professional decision makers, building experts and ordinary building users, it has to be easy to understand and simple to use. Data is collected through interviews with primary stakeholders, and there are no new calculations in the tool. An interviewer collects facts about the project beforehand and checks them with the stakeholders during the interviews. Interview questions are standardised with minor deviations, depending on stakeholders and building types. The evaluation of a project is based on subjective assessments, but also supported by project facts. Furthermore, there is a written explanation for a certain rating and the information about valuator is available too.

The RENO-EVALUE model is in Figure B10.1, illustrated as a radar diagram in which it is possible to rate parameters and their factors with grades 1–5 from low to high. For simplicity, only the results of one person's evaluation are shown, but the intention is that evaluations should be made by all the important stakeholders and compared. It is possible to make the rating before, during and after the energy renovation is completed, which in the end makes it possible to compare the expectations with the obtained results.

Benefits

The advantages of RENO-EVALUE are that it does not take a long time to make the evaluations, the graphical illustration of results is easy to understand and the model provides a quick overview of the current situation seen from a certain stakeholder's perspective. It can, for instance, be useful in the early phases of the energy renovation projects in order to improve the matching of expectations between different stakeholders and defining the success criteria for a project. After the project is completed, the evaluation results from the initial phase can be used to determine to which degree the success criteria are fulfilled. The evaluations can internally be used to compare before and after situations and externally for experience exchange and comparison between different projects.

Supplementary information about the project

The development of RENO-EVALUE was part of a larger European research project called ACES – 'A Concept for Promotion of Sustainable Retrofitting and Renovation in Early Stages', which was conducted in 2011–2013. DTU was responsible for the Danish part of the project with support from the Danish Energy Agency. The other partners in the project were Royal Institute of Technology (KTH) in Sweden and Frederick Research Centre on Cyprus. RENO-EVALUE was the primary contribution to the ACES project from DTU. Further information on the ACES project can be found on CFM's website: www.cfm.dtu.dk/english/research-projects/completed-projects/aces.

Literature guide

The chapter is mainly based on a report from the ACES project, which also includes an example of the use of the tool in a case study:

Jensen, Per Anker (ed.): *Economical and environmental benefits of restoration*. ACES project. Work Package WP 2. Joint Report. Centre for Facilities Management – Realdania Research, DTU Management Engineering, November 2013.

RENO-EVALUE is presented in full, including all four cases, in a report in Danish:

Jensen, Per Anker and Esmir Maslesa: *RENO-EVALUE – Et værktøj til målformulering og evaluering af bygningsrenovering (RENO-EVALUE – A tool for formulation of objectives and evaluation of building renovation)*. Research Report 8.2013. Centre for Facilities Management – Realdania Research, DTU Management Engineering, October 2013.

Other publications on RENO-EVALUE include the following conference paper and scientific article:

Jensen, Per Anker, Esmir Maslesa, Navid Gohardani, Folke Björk, Stratis Kanarachos and Paris A. Fokaides: Sustainability evaluation of retrofitting and renovation of buildings in early stages. *Proceedings of 7th Nordic Conference on Construction Economics and Organisation, NTNU, 12–14 June 2013*.

Jensen, Per Anker and Esmir Maslesa: Value based renovation: A tool for decision-making and evaluation. *Building and Environment*, No. 92, October 2015, pp. 1–9.

All reports from the ACES project can be found at www.cfm.dtu.dk/english/research-projects/completed-projects/aces/reporting

Use of the tool

As shown in the table that follows, the tool is particularly useful in two of the five processes presented in Part A: building project and process optimisation. The tool can be used as a decision support and evaluation tool, possibly in combination with usability briefing (see Chapter B.3). The tool was developed to be used for different building types, but it is, in particular, appropriate for renovation projects with a close collaboration between the client, users and FM organisation.

Process	Phase							
Strategy development	A	B	C	D	E	F		
Organisational design	A	B	C	D	E	F		
Space planning	A	B	C	D	E	F	G	H
Building project	A	B	C	D	E	F	G	H
Process optimisation	A	B	C	D	E	F		

For building projects, the tool can primarily be used in renovation projects. Here it can be used by managers and staff in client and FM functions and by their consultants through all phases as a basis for dialogue with other project partners and stakeholders, as well as in the definition of goals during the pre-project phases and in building briefings (phases A–C). Further, it can be used to conduct evaluations during the following project phases and the post-project phases (phase D-H).

For process optimisation, the tool can primarily be used in activities that involve changes in buildings. Here the tool can also be used by managers and staff in client and FM functions and their consultants through all phases from evaluation of current performance and improvement potentials to evaluation of new performance and assessment of need for further optimisation (phases A–F).

B.11 Energy-efficient FM

Christian Stenqvist

Introduction

One of the main challenges of reaching energy efficiency policy objectives is achieving major improvements in energy performance in the existing building stock, which may be operational for decades or centuries ahead. For a successful transition in which the market for energy efficiency services is a main impetus, several stakeholders and perspectives need to be involved. Firstly, stakeholders on the energy demand side – in this case building owners, FM organisations, users and residents – will have to be adequately informed and motivated to undertake investments in energy-efficient building renovations and operations. The public sector is expected to set an example. Secondly, to govern private- and public-sector decision-making, government policies need to internalise externalities related to energy use and effectively remove other obstacles to close the energy efficiency gap. Thirdly, providers of energy-efficient products and services need to be responsive to customer demands and create competitive offers and business models that can overcome persistent market barriers.

This chapter involves the three main stakeholder categories. It provides an assessment of a Swedish state policy for stimulating municipal-level energy efficiency strategies. It asks the question, how the demand and supply side of the energy efficiency service market can be better matched. Key processes are identified for the assessment of municipal FM organisations' in-house capabilities in energy-efficient FM (EEFM). As an output it, suggests an approach to engage municipal FM organisations on behalf of building owners and users, in dialogue and collaboration with external energy efficiency service providers. Following this approach may lead to better service design based on an in-depth understanding of customers' situations and preferences, strengths, weaknesses and improvement potentials in EEFM. The result is based on an interview survey with eight municipal FM organisations in the southern part of Sweden.

Strategies and models for energy efficiency improvement

As described in Chapter B.6, intended strategy can be different from realised strategy, which is relevant in the case of municipal agencies and their energy efficiency strategies. Elected politicians ought to make principal and strategic decisions about public services, while administrations and employees are responsible for preparing decisions and carrying out operations. Municipal FM organisations are relatively large and divided into administrations and municipally owned companies with different areas of responsibility, and with contractors and consultants hired for various tasks.

To examine the links between politically endorsed strategies and actual operations, respondents were asked about four proposed models that describe the energy efficiency practices of the FM organisations. The models, shown in Figure B11.1, were derived from a study of energy efficiency practices in Danish municipal buildings and then further developed and adapted to the Swedish context (see literature guide). The horizontal axis distinguishes between the involvements of internal or external resources, whereas the vertical axis distinguishes between an incremental or more direct rate of energy efficiency improvement.

In most cases, the answers about utilised models relates only to non-residential buildings. Optimisation of operations (A) and maintenance and energy efficiency improvement (C) were often stated as applied and preferred models. On the other hand, some interviews revealed scepticism and disappointment related to the ESCO-solution (B) – e.g. fails to deliver expected energy savings and negatively influence other performance criteria, overly focussed on least cost measures, troubled by technological and contractual lock-in effects. ESCO is treated in more detail in Chapter B.9.

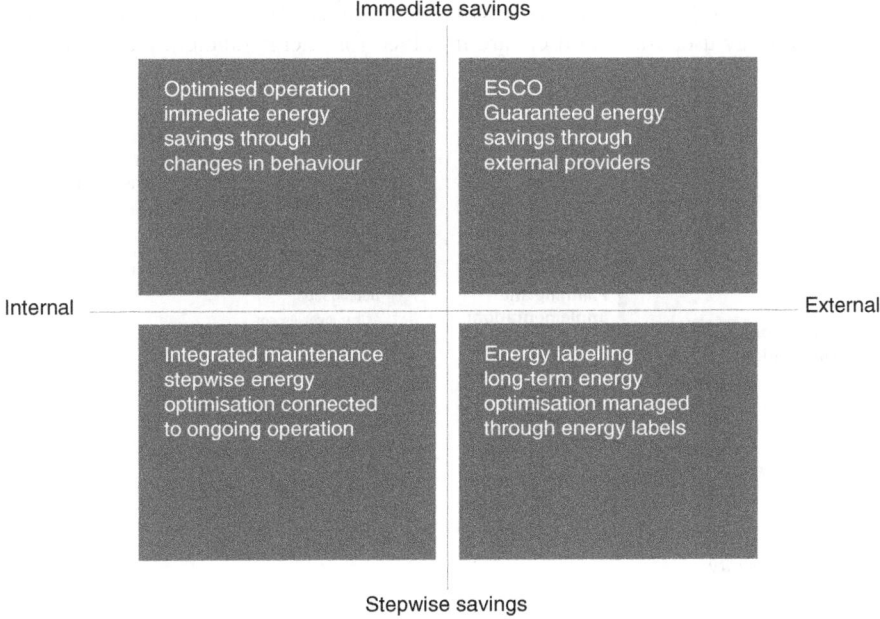

Figure B11.1 Four models for energy efficiency in municipal buildings

Developing energy-efficient FM

In search of the capabilities of EEFM, the interviews probed the concept to identify phases and key processes. As demonstrated by Figure B11.2, the phases form an energy efficiency value staircase starting from basic orientation about problems and solutions and extending up to demonstration of results from implemented measures. Each phase consists of the underlying key processes that, based on the interviews, are suggested as important constituents of EEFM.

For each key process, the municipal FM organisation can be assessed against a descriptive yardstick and assigned a maturity level – e.g. from 'ignorant' to 'professional'. Awareness raising about the current situation and improvement potentials should support decisions about steps to take in-house or in collaboration with energy efficiency service providers.

The content of the four phases are described as follows.

> *Phase 1 – Orientation:* Develop awareness of problems related to excessive energy use (e.g. economic, environmental and social) and start to perceive energy efficiency as a viable and cost-effective solution. Make data and information accessible about buildings' energy use, energy baselines and energy efficiency potentials.
>
> *Phase 2 – Planning and implementation:* Procedures are developed to ensure that procurement and project planning delivers energy-efficient equipment and services. Financing solutions are identified and secured, preparations are made for installations and measures on energy using equipment.
>
> *Phase 3 – Operation and supervision:* Technical operations are made to optimise energy use, and end users are involved for energy efficiency improvement

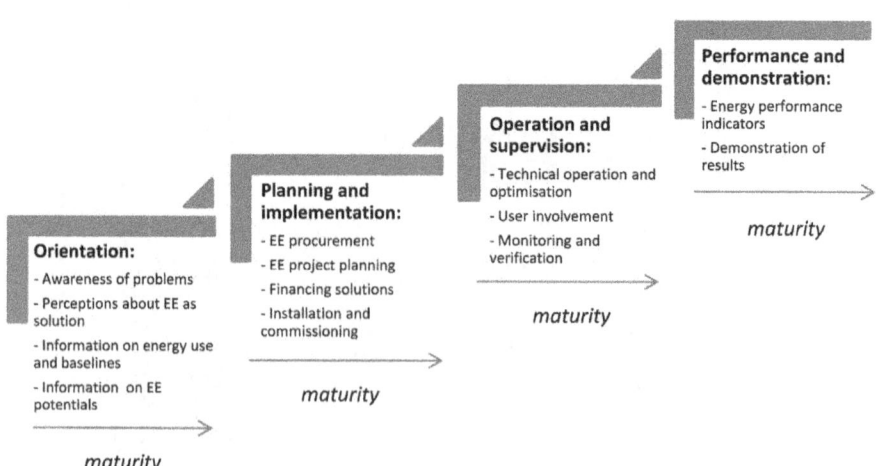

Figure B11.2 A value staircase with phases and key processes of municipal energy-efficient FM

via behavioural measures. Measurement and verification are used to follow up energy-efficiency measures and results.

Phase 4 – Performance and demonstration: Indicators are used to demonstrate the energy performance development for aggregated building stock or certain building types. Energy efficiency results are also translated and communicated in terms of multiple benefits – e.g. better indoor climate and improved environmental performance.

Users of the model can be municipal FM organisations that want to evaluate and improve their internal capabilities within energy-efficient FM. Private and external energy efficiency service providers can also use the model as a tool to initiate dialogue with potential customers (i.e. municipal FM organisations) and to develop tailor-made offers that generate customer benefits. The complete model, the maturity matrix for energy-efficient FM, has been tested by both target groups. According to the project's business partner, the Swedish consulting company EVU Energi & VVS Utveckling, it contributed to additional sales. A focus group validation was conducted during a meeting for municipal strategists, where they self-diagnosed their organisational maturity in EEFM, which generated positive feedback.

Literature guide

The chapter is based on the following conference paper:

Stenqvist, Christian, Susanne Balslev Nielsen and Per-Otto Bengtsson: Dialogue and collaboration for Energy-efficient FM: Public sector strategies and the role of external service provider. Paper in K. Alexander and I. Price (eds.): *People Make Facilities Management.* EuroFM Research Papers 2015. 12th EuroFM Research Symposium, Glasgow, 1–3 June 2015.

Figure B11.1 is adapted from a model presented in the following article:

Jensen, Jesper Ole, Ole Michael Jensen and Dorte Nørregaard Larsen: Modeller for energibesparelser i kommunale bygninger (Models for energy savings in municipal buildings), Danish Building Research Institute, SBi, Aalborg Universitet, Copenhagen, 2013.

The results of the project are documented in a report in Swedish:

Stenqvist, Christian: *Energieffektivisering i skånske kommuner, fastighetsförvaltningar och bolag.* Lund University Publications, 2015, available at https://lup.lub.lu.se/search/publication/7456369#?

In addition, a scientific article has recently been published from the project:

Stenqvist, Christian, Susanne Balslev Nielsen and Per-Otto Bengtsson: *A tool for sourcing sustainable building renovation: The energy efficiency maturity matrix.* Sustainability, No. 10, 2018.

Use of the model – assessed by Per Anker Jensen

As shown in the table that follows, the model is particularly suitable for use in two of the five processes presented in Part A: organisational design and process optimisation. The model can be used as an analysis tool and decision support in connection to sourcing and preparing collaborative relationships. The model can, with benefits, be used in combination with the method in Chapter B.9 and the model in Chapter B.27.

Process	Phase							
Strategy development	A	B	C	D	E	F		
Organisational design	A	B	C	D	E	F		
Space planning	A	B	C	D	E	F	G	H
Building project	A	B	C	D	E	F	G	H
Process optimisation	A	B	C	D	E	F		

For organisational design, the model can mainly be used by managers and staff at the strategic level to analyse and decide on sourcing and possibly establishing collaboration with external providers in the initial phases to identify the need for new knowledge and competences (phases A–D).

For process optimisation, the model can mainly be used by managers of internal FM functions to plan innovation processes from the initial phase with evaluating the current performance and improvement potential to evaluating new FM performance (phases A–D).

B.12 Improving the environmental performance of buildings

Esmir Maslesa

Introduction

Benchmarking of environmental building performance and mapping of the underlying processes through relevant IT systems in FM has had its challenges in practice. Partly because of missing or incorrect data about building operation, including consumption data for electricity, water and heating, but also due to the obscurity of the underlying processes for consumption settlement and benchmarking of environmental performance in commercial and public FM organisations.

Nowadays, there are IT systems such as Integrated Workplace Management Systems (IWMS) and Energy Management Systems (EMS) that can monitor and report on environmental building performance. This development provides new opportunities to base environmental KPIs on actual consumption data and opens opportunities for further performance optimisation and scenario planning. Improving environmental building performance plays an important role in the sustainable transition, and, therefore, there is a need to study the effects of IWMS and EMS implementation for environmental building performance in FM organisations.

Purpose

To be able to improve environmental performance through IT systems like IWMS and EMS, several conditions must be realised. One of them is that relevant environmental categories are identified and converted into measurable KPIs that can be monitored and analysed in IT systems. A systematic literature review of 68 research articles in 2016 has identified eight environmental categories and found examples of KPIs that can be used for monitoring and benchmarking of environmental performance. The study results are briefly presented in this chapter.

Target group

The target group for the use of identified environmental categories and their KPIs are stakeholders, such as property owners, real estate managers and facility managers, who aim to visualise and improve the environmental performance of their property portfolio using IT systems like IWMS and EMS.

Environmental building performance

Once buildings are built, it is difficult to change decisions made in relation to their project design, such as building orientation, window/wall surface ratio, and location of technical installations. Environmental building performance focusses on the buildings' environmental properties under actual operating conditions. Environmental building performance depends on several factors, such as building design, choice of building materials, building location and usage patterns. The literature study from 2016 shows that the processes during building operation and maintenance have much higher financial costs and environmental impacts than for the design and construction stage. For example, the findings show that 80%–90% of negative environmental effects occur during the building's use stage, while 10%–20% of the environmental impacts relate to the manufacturing of building materials and the construction stage. The study also points out that environmental building performance can be improved with appropriate operational and maintenance activities, such as ongoing maintenance, building renovation and optimised operating hours.

The optimisation potential can be visualised through continuous collection, visualisation and analysis of actual performance data on various environmental categories. In relation to this, the literature study has identified eight environmental categories that can be monitored and benchmarked to improve environmental performance. Table B12.1 shows the eight environmental categories as well as some examples of their KPIs.

Table B12.1 Environmental categories and examples of their KPIs

Category	Examples of KPIs
Energy management	Energy consumption (kWh, MWh, GJ) Energy saving potential (kWh, MWh, GJ or %) Energy supply (renewable energy (%), non-renewable energy (%))
Emissions	'CO$_2$, CO$_{2e}$, NOx, SO$_2$, etc.'
Water management	Water consumption (m^3) Water saving potential (m^3 or %) Water supply: local water, rainwater (m^3 or %) Water pollution
Waste management	Daily waste (kg, t) Construction waste (kg, t) (manufacturing, treatment, disposal)
Space management	Property site (m^2) Total building area (m^2) Building capacity (m^2/person), Vacant/occupied area, common area (%)
Building materials	Type (concrete, wood, glazing, etc.) Durability (years) Thermal properties (U-value) Maintenance needs (frequency)

Category	Examples of KPIs
Indoor environmental quality	Thermal comfort (°C)
	Humidity (%)
	Daylight factor
	Air quality/air pollution (ppm)
Recycling	Building materials (concrete, wood, glazing, etc.)
	Building components (doors, windows, etc.)

Use of environmental KPIs in research and practice

The study has revealed that the identified environmental categories and their KPIs are unequally represented in the research and that energy management and emissions are especially in focus. During buildings' use stage, most of the focus is on the electricity and heating consumption, energy saving potentials and greenhouse gas emissions induced as a direct consequence of energy consumption. Moreover, the focus is also on water consumption and waste management, while space management is especially relevant for commercial buildings. On the other hand, there is a lack of research on the importance of building materials, indoor environmental quality (IEQ) and recycling potentials for environmental performance during building use stage.

Similar trends are also observed in practice. Currently, the Danish Building and Property Agency (BYGST) is implementing IT systems IWMS and EMS, and in relation to this, it has been decided that the systems must be able to monitor and benchmark over 1,800 leases on electricity, water and heating consumption. The EMS will be used for collection, visualisation and analysis of energy data, which subsequently will be interfaced with building data from IWMS so that BYGST becomes able to benchmark electricity, water and heating consumption per area unit (m^2) for each lease, and among the various customer groups. Moreover, BYGST's customers/tenants and associated facilities managers will be able to monitor their energy consumption at an hourly level through the web-service solution, as presented in Figure B12.1.

In general, KPIs relating to energy management, greenhouse gas emissions and water management are mostly applied in both FM research and practice, while KPIs on IEQ, building materials and recycling potentials are often neglected and lack detailed data.

The indoor environmental quality has a large effect on energy consumption and thus environmental building performance, but its KPIs are often overlooked, both in research and in practice. In the literature, the IEQ is typically considered a social category. However, the IEQ is here considered to be an environmental category due to its significant importance for energy consumption. Indoor environmental quality KPIs, such as indoor/outdoor temperature, thermal comfort, relative humidity and indoor air quality, are of great importance to energy consumption, but their data often does not exist or is stored in single discipline

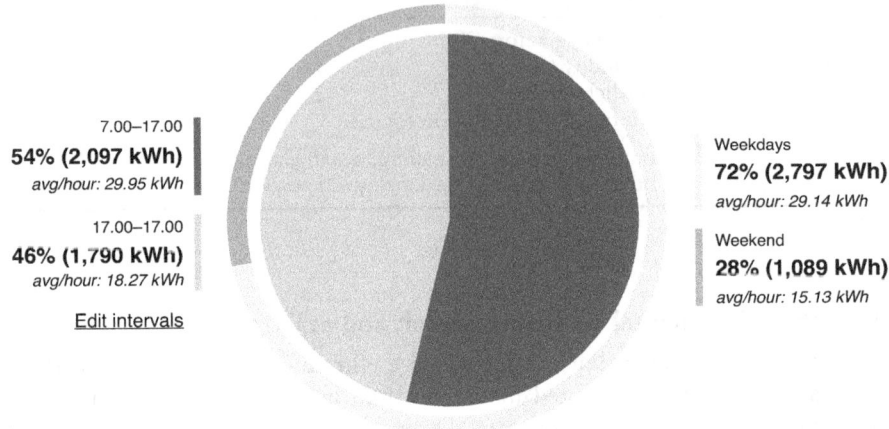

7.00–17.00
54% (2,097 kWh)
avg/hour: 29.95 kWh

17.00–17.00
46% (1,790 kWh)
avg/hour: 18.27 kWh

Edit intervals

Weekdays
72% (2,797 kWh)
avg/hour: 29.14 kWh

Weekend
28% (1,089 kWh)
avg/hour: 15.13 kWh

3,887 kWh

Standby 14 kW pr. hour*
The calculation is based on the period June 4, 2018–June 11, 2018

Figure B12.1 An example of electricity consumption monitoring in one of BYGST's office leases. The figure shows weekly electricity consumption during working hours (7:00–17:00) and outside of working hours, as well as the distribution of electricity consumption between weekdays and weekend. The figure shows that 72% of consumption occurs during weekdays and 28% during weekend. The figure also shows that 46% of electricity consumption occurs outside of working hours and 54% during working hours. The data is collected through EMS and is available online for the tenant through web-service solution

Source: Webtools.

systems like BMS that typically are not interfaced with other FM systems, resulting in a loss of important information on IEQ performance.

Benefits and recommendations when using KPIs

One of the benefits of IT systems like IWMS and EMS is that they quickly can show actual energy consumption through various KPIs. An EMS can collect hourly data directly from the consumption metres and be configured to visualise and analyse the consumption data so that different user groups like facilities managers, energy managers and tenants can easily monitor actual electricity, water and heating consumption. Validated consumption data can subsequently be combined with the building's master data coming from IWMS, thus enabling development of additional KPIs, such as consumption/area or consumption/person.

These additional KPIs can provide deeper insight into environmental building performance of each building and allow benchmarking across property portfolio based on common parameters.

KPIs can be used to monitor and control desired effects. The selection of KPIs and their setup is of importance to the end result, so it is essential to carefully consider which KPIs are required to use and for what purpose. It is recommended to use a limited number of KPIs for each of the eight environmental categories rather than having several KPIs covering a few specific environmental categories. A limited number of KPIs for each environmental category makes it easier to keep an overview of the holistic evaluation of environmental building performance.

Many KPIs are often a combination of several parameters. For example, the KPI for building capacity is a combination of the building area and number of occupants (m^2/person). Therefore, it is a prerequisite that both parameters contain validated data to make this KPI useful. In addition, the use of KPIs with standardised measurements (e.g. reading frequency: hourly, daily, weekly, monthly) and units of measurement (m^2, kWh, m^3, etc.) is recommended so that indicators can be compared to targets, milestones and across the property portfolio. It is important that the collected data used in KPIs is valid and reflects the actual conditions of the building in order to get a reliable result of current performance.

Supplementary information on the project

The Danish IT company KMD has started a new business area on IT systems within facilities management in collaboration with the software developer Trimble (former Manhattan Software Group, UK). Relating to this, KMD has recently started to provide EMS solutions to their customers. Since it is a new business area for KMD, the company has decided to support an industrial PhD project within this field to deliver valuable knowledge about the implementation and further development of IWMS product 'KMD Atrium' (Manhattan software). Through case studies of IWMS and EMS implementations, this industrial PhD provides new knowledge about IT needs in public and private real estate organisations and gives KMD unique insights into the strengths and weaknesses of IWMS and EMS regarding real estate management and environmental building performance. The PhD study started in 2016 and will be finished in 2019.

Literature guide

The chapter is based on the following conference paper and scientific article:

Maslesa, Esmir, Susanne Balslev Nielsen, Morten Birkved and Jannik Hultén: Environmental indicators for non-residential buildings: When, what, and how to measure? Paper in Susanne Balslev Nielsen, Per Anker Jensen and Rikke Brinkø (eds.): *Research Papers for EuroFM's 16th Research Symposium at EFMC2017, 25–28 April in Madrid, Spain*. EuroFM, Centre for Facilities

Management – Realdania Research, DTU Management Engineering, and Polyteknisk Forlag, April 2017.

Maslesa, Esmir, Per Anker Jensen and Morten Birkved: Indicators for quantifying environmental building performance: A systematic literature review. *Journal of Building Engineering*, Vol. 19, pp. 552–560.

The implementation of IT systems for environmental building performance is treated further in the following conference paper:

Maslesa, Esmir and Per Anker Jensen: The implementation impacts of IT systems on energy management in real estate organisations. Paper in Matthew Tucker (eds.): *Research Papers for EuroFM's 17th Research Symposium at EFMC2018, 5–8 June 2018 in Sofia, Bulgaria.* EuroFM, 2018.

Environmental building performance is treated further in the following scientific article:

Jensen, Per Anker, Esmir Maslesa, Jakob Brinkø Berg, Christian Thuesen: 10 questions concerning sustainable building renovation. *Building and Environment*, Vol. 143, 2018, pp. 130–137.

Use of the tool – assessed by Per Anker Jensen

As shown in the table that follows, the tool is particularly useful in one of the five processes presented in Part A: process optimisation. The tool can be used as decision support in the selection of KPI's for environmental building performance and as a basis for using such KPI's to optimise performance.

Process	Phase							
Strategy development	A	B	C	D	E	F		
Organisational design	A	B	C	D	E	F		
Space planning	A	B	C	D	E	F	G	H
Building project	A	B	C	D	E	F	G	H
Process optimisation	A	B	C	D	E	F		

For process optimisation, the tool can be used by managers and staff responsible for sustainability in FM organisations through all phases from evaluation of current performance and optimisation potential to assessment of need for further optimisation (phases A–F).

B.III

Innovation and partnerships

B.13 ICT in the FM supply chain

Ada Scupola

Introduction and background

One of CFM's projects at Roskilde University concerned 'ICT-based innovation in the FM supply chain'. As part of this study, an interview survey was conducted among Danish companies in 2008–2009 regarding the use of information and communication technology (ICT) within FM. The purpose of the interviews was to identify how the introduction and expansion of ICT systems within the FM supply chain take place, and what factors promote and hinder such innovations.

As a basis for the study, a literature review of previous research was carried out within the following three research areas: supply chain management (SCM), Facilities Management (FM) and innovation theory. The literature review focussed on the introduction and diffusion of ICT in supply chains. A result of the literature study was the formulation of the model shown in Figure B13.1 of the most important factors affecting the introduction of ICT in the supply chain in FM.

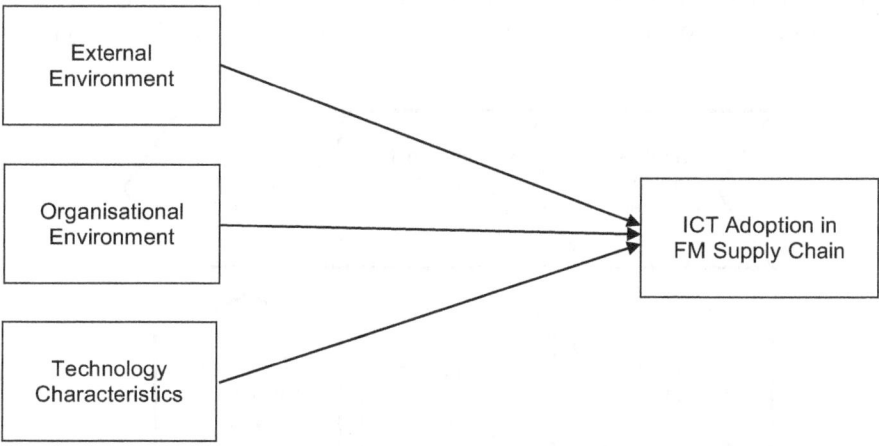

Figure B13.1 Model for key factors of importance for ICT implementation in the FM supply chain

In relation to the external environment, factors such as customer/provider dependence, competitive intensity and intensity of IT activities are included in the model. Regarding the organisational environment, management's IT knowledge includes the degree of centralisation and formalisation of the IT departmental structure. The most important technological characteristics include relative benefits, compatibility and complexity. The interview survey took the starting point in these theoretically based factors, and the importance of such factors as inhibitors and promoters was investigated based on the interviewees' statements.

Execution of the study

The interviews were conducted in late 2008 and the first half of 2009. Twelve companies participated in the study. The companies were selected in the following way: five of the companies were FM users – hereinafter referred to as customer companies – and included both private and public companies; two of the companies had a large in-house FM function, while the three others had predominantly outsourced their facilities services. In addition, three companies were providers of facilities services – hereinafter referred to as FM providers. The remaining four companies included two suppliers of ICT systems – hereinafter referred to as ICT suppliers – and two consultancy companies.

Figure B13.2 illustrates the FM supply chain, which again illustrates the interrelationship between the different types of companies. For customer companies, there is typically a distinction between core activities, as an in-house FM function is a supplier, while external FM providers are suppliers to the customer's in-house FM function – possibly in the form of a small FM contract management unit. The aforementioned represents the horizontal supply chain in the figure. ICT suppliers and consultancy companies have a more flexible position in relation to the FM supply chain, as they can both provide services to the customer's core activities and to internal FM functions and FM providers. In addition, often their services are not provided on a regular basis, as are the services in the horizontal supply

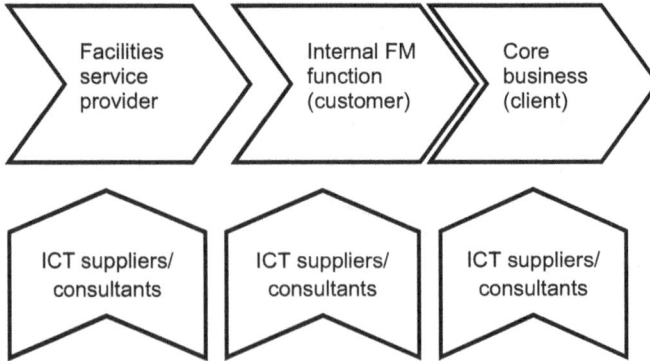

Figure B13.2 Companies in the FM supply chain

Table 13.1 IT FM systems represented in the interview survey

IT FM system	Description
Byggeweb	Danish web-based, project-web system for storing and exchange of documentation about building projects and existing buildings. Developed and introduced in 1997 by the company Byggeweb A/S, which also has developed CoreFM.
Caretaker	Danish FM system with a main focus on O&M. Developed and introduced in 1994 by the consulting company COWI. Caretaker is used by many Danish municipalities.
CoreFM	Danish CAFM system with a main focus on space management. Developed and introduced in 2004 by Byggeweb A/S.
DriftsChefen	Icelandic FM system with a main focus on O&M. Developed and introduced in 1995 by the consulting company ICEconsult.
EAM	International asset management system with a main focus on maintenance. EAM is marketed internationally by the IT company Infor.
FM Anywhere	Danish FM systems with a main focus on O&M. Developed and introduced in 2004 by the consulting KeyCon.
Maximo	International maintenance system from 1984 marketed by IBM.
Navision	International ERP system marketed by Microsoft. A version called Navision Stat is targeted towards state institutions.
SAP	International ERP system developed and marketed by the German company SAP AG.
Vista FM	International FM system also called TAC FM with a main focus on O&M. Developed in 1991 by the Finnish consulting company Granlund and marketed internationally by the building automation company TAC – part of Schneider Electric.

chain. They are, therefore, illustrated as a series of parallel vertical delivery chains in the figure. The role of different actors in relation to FM processes is treated in more detail in Chapter B.15.

Table B13.1 provides an overview of the ten different IT FM systems that were represented in the interview survey – excluding Excel and similar generally applicable applications, which are probably the most widely used in FM.

Main results

The study showed that there is a big unexplored potential for using ICT in the FM supply chain. One of the interviewees estimated that only 20% of companies who could use ICT in FM did so in an efficient manner. There is thus a large potential market for ICT in the FM area and the possibility of major efficiency improvements. Among the barriers to the introduction of ICT, the characteristics of technology are of significant importance. The systems themselves are not sufficiently user friendly, and across systems, there is insufficient compatibility for systems to exchange data and work together in a simple and flexible way. In addition, there are no common standards for data formats and classifications of data objects to make it easy to import and export data from and to the systems.

In relation to construction projects, construction and industry managers lack knowledge and sufficient attention to what kind of data is needed during the operational phase. Similarly, it is difficult to specify data requirements – including which type of data, when to deliver them and how to deliver and use them – in connection with the tendering process of FM projects. For providers, the lack of standardisation and compatibility means that they may have to work with different ICT tools for different customers.

One important result of the study is that outsourcing can help promote the use of ICT systems. This applies particularly for public customer companies, which are bound to tender at intervals of a few years, but also for private customer companies, as switching providers represents a significant risk for the customer of losing data that is located in the provider's ICT systems. Therefore, outsourcing provides a strong incentive for customer companies to establish their own ICT systems to gain control over data about one's own facilities, regardless of whether that data is primarily used by one's own employees or FM providers.

It is clear that public regulation, not least in terms of the requirements of the digital construction initiative, has helped to promote the introduction of ICT systems, especially between the state building and operating authorities. A stronger management focus on FM is also advancing the introduction of ICT systems, as well as the developments in continuation of the structural reform, with the establishment of central property units in municipalities, have also promoted developments in the ICT field. Among the arguments for introducing ICT systems is better overview of property portfolios, better short and long-term planning of FM activities and resource needs, better financial management and better systematic information processing and service delivery. Conversely, lack of employee resources and competencies, as well as the need to implement organisational adjustments in connection with the introduction of ICT systems represent organisational factors that may constitute barriers to the spread of ICT in FM.

Conclusion

Although this study is of limited scope, it provides insights into some of the typical conditions that affect the introduction of ICT in FM. It points out a number of well-known factors concerning standards, data classification and systems compatibility. The identified advantages of introducing ICT systems in FM and the organisational challenges that the introduction of such new technology implies is also well known.

One of the most significant new factors that the study points out is that outsourcing of FM can actually give customer companies new incentives to self-establish ICT systems to maintain data about their own facilities. This can have a significant effect on the way customer and provider companies collaborate on ICT systems, as well as on how they use and exchange of data. In addition, it could potentially open up a larger market of ICT providers offering to maintain and operate systems thus operating as application service providers for customer companies regardless of who is actually the FM provider.

Literature guide

The chapter is based on the following research report:

Scupola, Ada and Per Anker Jensen: *Information and communication technologies and supply chain in facilities management in Denmark*. Centre for Service Studies, Research Report 09:3, and CFM, December 2009.

Results from the study have also been presented in the following conference papers and scientific articles:

Jensen, Per Anker and Ada Scupola: ICT adoption in the Danish facilities management supply chain: What are the factors that matter? Construction Matters Conference, Copenhagen, 5–7 May 2010.

Scupola, Ada: The relation between innovation sources and ICT roles in Facility Management Organizations. *Journal of Facilities Management*, Vol. 12, No. 4, 2014.

Scupola, Ada: ICT adoption in facilities management supply chain: The case of Denmark. *Journal of Global Information Technology Management*, Vol. 15, No. 1, 2012.

Scupola, Ada: Relationship between outsourcing and ICT adoption in facility management supply chain. Mediterranean Conference on Information Systems, MCIS 2010, Tel-Aviv, Israel, 12–14 September 2010.

Scupola, Ada: Information and communication technologies and supply chain: lessons from the Danish facilities management service sector. Global Information Technology Conference – GITC, Washington, DC, 19–22 June 2010.

Scupola, Ada: Benefits and barriers of information and communication technologies adoption in facilities management services supply chain. European Academy of Management – EURAM, Rome, 19–22 May 2010.

Use of the model – assessed by Per Anker Jensen

As shown in the table that follows, the model is particularly suitable for use in one of the five processes presented in Part A: process optimisation. The model can be used as a decision support tool in consideration of implementing ICT systems in FM, possibly in combination with the method in Chapter B.28.

Process	Phase							
Strategy development	A	B	C	D	E	F		
Organisational design	A	B	C	D	E	F		
Space planning	A	B	C	D	E	F	G	H
Building project	A	B	C	D	E	F	G	H
Process optimisation	A	B	C	D	E	F		

For process optimisation, the model can mainly be used by managers and staff on the strategic and tactical levels in FM functions in the initial phases to evaluate current performance and improvement potential and identify, decide and implement changes (phases A–C).

B.14 Drivers for innovation in services

*Per Anker Jensen (based on work
by Giulia Nardelli)*

Introduction

In the literature on innovation in services, it is often taken for granted that inter-actions between stakeholders occurs as collaboration without any friction. It is neglected that tensions and conflicts emerge during the innovation processes, but to present the interplay between the parties as based on uncomplicated collabo-ration is dangerous, because it creates a distorted picture of reality. This causes misunderstandings among practitioners, but it also limits the theoretical under-standing of innovation practice.

To go beyond this limitation in the existing literature, the study presented in this chapter has through longitudinal case studies investigated how and why ten-sions and potential conflicts between different stakeholders unfolds during inno-vation processes. By taking the tensions and conflicts between the stakeholders as the focus of the analysis, the ongoing changes in the relationship between the development of such conflicts and the development in the innovation process over time were followed.

The dialectic process model

A recurring finding in the collected data is that, when a change, caused either by an exogenous shock or an endogenous decision, is introduced into the system, it challenges the status quo – i.e., the balance between needs and expectations of different stakeholders that was previously achieved. As a consequence, each set of stakeholders needs to deal with issues, needs and expectations that might be very different from their own, which, in turn, causes tensions between parties. Such ten-sions trigger a dialectic motor of change, which in the study was named *stakeholder dialectics*. Stakeholder dialectics is defined as a constructive mode of change that takes place within a network of two or more stakeholders. Constructive mode of change refers to the conflict between the thesis and anti-thesis, which eventually resolves in a synthesis. The resolution of the conflict – i.e., the synthesis – generates a break with the past basic assumptions that regulate their relationships.

A main result of the study was the proposal of a process model of innovation in services, which is centred on the dialectic motor of change and driven by

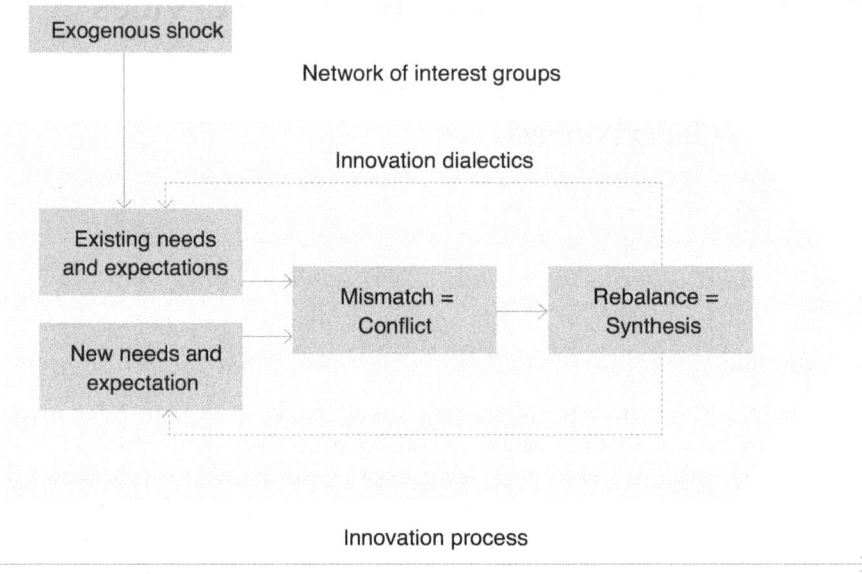

Figure B14.1 The dialectic process model for service innovation

stakeholder dialectics, as illustrated in Figure B14.1. In the figure, thesis and anti-thesis are associated, respectively, to (1) new needs and expectations of one (or more) sets of stakeholders, originated by the introduction of a change into the system, and (2) existing needs and expectations of the other stakeholders. The conflict is the mismatch of needs and expectations resulting from the confrontation of diverse stakeholders, which eventually resolves in a synthesis.

In short, when change is introduced into the system, stakeholder dialectics kick in as new needs and expectations (of one or more sets of stakeholders) are confronted with existing ones. The resulting mismatch of needs and expectations turns into an explicit or implicit conflict, which is resolved by re-balancing the interests of the different parties into the synthesis. Finally, the synthesis feeds back into the process as (1) the new status quo for some stakeholders (dashed line in the figure) and as (2) endogenous change for others (dotted line in the figure). In the model, stakeholder dialectics are represented as a big arrow, as they are intended as a transition and not a status. The succession and combination of various episodes of stakeholder dialectics are what constitute the innovation process and contribute to eventually reaching innovation outcomes. In other words, the proposed model of innovation in services stresses the mismatch of needs and expectation of different stakeholders as one of the driving forces of innovation in services.

Case on Novozymes

The model was developed partly from a case study of the development of FM in Novozymes (NZ), a Danish, multinational organisation (6,200+ employees). Main results from this study will be presented in the following.

NZ was established in the year 2000 as a new company separated from the pharmaceutical corporation Novo Nordisk. The core business of NZ lies within industrial biotechnology, with a strong focus on enzyme production. The set of interest groups under investigation includes (1) the developing internal facility service unit, heretofore also referred to as NZ FM; (2) the organisation, heretofore also referred to as NZ, which is supported by such unit; (3) its employees, who are served by NZ FM; and (4) the outsourced providers.

Since 2000, the intertwining of tensions and conflicts between NZ's interest groups and the innovation processes carried out in their network developed over time and went through four critical phases or temporal brackets: (1) a de-merge crisis, (2) a financial crisis, (3) a global shift and (4) an organisational change.

The de-merge crisis

When NZ de-merged from its mother company, it shortly realised that it needed to determine how to deal with facility services, which were previously taken care of by the facility service unit of the mother company. An embryonic FM unit was created, composed of only a director from the purchasing unit, who could initially only dedicate 20% of his time to ensure that facility services were allocated efficiently to the employees. Soon after the de-merge, he had to re-negotiate the contract with the outsourced provider, and together with NZ's executive management, he decided to discontinue the existing relationship to obtain better conditions, such as greater transparency, cost competitiveness and better services.

A major conflict arose as the decision to discontinue the relationship caused the refusal by the former external facility service provider to share any information and/or data about the past facility service provision with the embryonic facility service unit. This meant that the latter had to start from scratch to determine how to ensure that employees could carry out the activities related to the core business without noticing the ongoing shift of providers. This conflict caused the threat of a mismatch between the needs and expectations of NZ and its employees, and the needs and expectations of the embryonic facility service unit and thus of a potential conflict between interest groups.

The solution for NZ FM was to find new facility service providers. This was not part of a strategy – it was damage control. In addition, significant effort was invested in designing and implementing a communication strategy that would explain the ongoing changes to NZ's employees, thereby limiting dissatisfaction and related tensions with the newly appointed internal and external facility service providers.

The financial crisis

Once the initial challenges were faced and external providers selected, NZ FM became an actual independent unit reporting to the vice president of stakeholder relations. The director from purchasing, who led the process from the time of the de-merger, was appointed director of the FM unit and new managers were hired. The newly formed facility service unit was responsible for a limited amount of facility services (real estate, technical maintenance and renovation, cleaning and catering, logistics) at NZ's headquarters in Denmark, while the other FM services (and those in other NZ sites around the world) continued to be managed by 'regular' employees within the local units, based on specific needs and personal preferences.

The more NZ FM developed, the more sophisticated the needs and expectations of the organisation as a whole and of the employees with regard to facility services became; they moved from the operational to the tactical and strategic levels. While an increasing number of facility services were assigned to the NZ FM unit, an exogenous shock hit NZ like many other companies – the financial crisis. This caused the executive management of NZ to require that the budget for service provision to the facility unit be reduced. This resulted in, among other things, space usage becoming important to focus on as a possibility to save cost. The exogenous shock generated a mismatch between the expectations of the employees, who were used to high-level services, and the needs of the executive management, which turned its attention to facility services as a potential source of cost savings. One of the side projects, which originated in connection to the financial crisis, was the reduction of travel expenses, and due to the financial crisis, a new travel policy was implemented. The NZ FM unit, therefore, became involved in the development and implementation of video conference rooms, which represented an innovation in NZ.

Global change

By 2009, NZ FM was composed of a team of facility service managers led by a facility service director who operated at the Danish level and reported to the vice president of stakeholder relations. Other than managing facility services in the best possible interest of the employees, NZ FM started working on the development of some 'transparency tools' that would ease communication with executive management. The goal of transparency had been one of the major drivers of NZ FM development since the time of the de-merger, with the aim of achieving a better quality/cost ratio for facility services. In addition, transparency would support better communication with the executive management, which in turn would result in increased awareness of the potential contribution that facility services could offer to the core business. The idea of extending the responsibilities of NZ FM outside the Danish sites and building a facility service unit to manage facility services on a larger scale (in other NZ sites around the world) was initiated once the executive management realised the potential of proper management of facility services. With the aid of consultants and academic facility service research, an analysis project called FM Deep Dive was conducted with ten other Danish multinational companies to explore how facility services could be managed. Two main dimensions were investigated: (1) centralisation and (2) globalisation.

FM Deep Dive provided the inspiration for a globalisation project, which was called Global FM (GFM), initiated in 2011. The GFM project was launched to identify the similarities and differences between facility services in Denmark and those around the world. The emphasis was not only on the facility services themselves but also on the needs and expectations of local employees and executive management, along with the cultural differences in people's behaviour, rules and regulations. The plan for a global organisation carried the risk of creating an imbalance between the satisfaction of local employees and the goals and objectives of the centralised management. Such risk was dealt with by creating a team of internal facility service managers from the different sites interested in the project to map local needs and expectations and compare them with the potential global requirements and standards.

Organisational change

In April 2013, the CEO, who had led NZ since the 1990s, retired, and the new CEO introduced several organisational changes immediately. Some of these changes had a direct effect on NZ FM and the provision of facility services. First, the new organisational structure put NZ FM together with all other support services under the responsibility of the vice president of global business services. Second, the facility service unit was divided into two entities led by two facility service directors: one responsible for Denmark (DK FM) and one for the Rest-of-the-World (ROTW FM). The goals and objectives of the GFM team were transferred to the newly formed ROTW FM unit, which basically had to define the scope of its service provision from scratch. The organisational change thus created a mismatch between the needs and expectations of the executive management and the newly reorganised facility service institution and its employees.

Literature guide

The chapter is mainly based on and treated in more depth in the following scientific article:

Nardelli, Giulia: Innovation dialectics: An extended process perspective on innovation in services. *The Service Industries Journal*, Vol. 37, No 1, 2017, pp. 35–56.

The topic is also treated in Giulia's PhD thesis:

Nardelli, Giulia: Innovation in services and stakeholder interactions: cases from facilities management. PhD Dissertation. Roskilde University. November 2014.

Further information about Giulia's research and publications can be found in Chapters B.15 and B.16.

Use of the model – assessed Per Anker Jensen

As shown in the table that follows, the method is particularly useful in two of the five processes presented in Part A: strategy development and process optimisation. The model can be used as a frame of understanding and analysis tool in management of service innovation. It can benefit from being used in combination with the method in Chapter B.15 and B.16. The model can also be used in relation to stakeholder management, possibly in combination with Value-Adding Management (see Chapter B.27).

Process	Phase							
Strategy development	A	B	C	D	E	F		
Organisational design	A	B	C	D	E	F		
Space planning	A	B	C	D	E	F	G	H
Building project	A	B	C	D	E	F	G	H
Process optimisation	A	B	C	D	E	F		

For strategy development, the model can be used by managers and staff, particularly on the strategic level in FM functions, to understand and analyse where we are now and contribute to defining strategy goals and develop, implement and re-assess strategy plans (phases B–F).

For process optimisation, the model can be used by managers and staff in FM functions and their consultants in the initial phases with evaluation of current performance and improvement potential and identification, decision and implementation of changes (phases A–C).

B.15 Management of innovation in FM

*Per Anker Jensen (based on work
by Giulia Nardelli)*

Introduction

In FM, just as in any other sector of the contemporary economy, innovation has become a not only recommended but also required element of survival and growth. External FM providers as well as internal FM departments are aware of the role of innovation as a tool to succeed and compete in the dynamic contemporary FM market. Some FM organisations have developed a dedicated innovation strategy and manage innovation and improvement strategy systematically. However, innovation and improvement processes are not just about creative ideas and project management, and many FM practitioners admit that they still struggle to establish innovation routines and to manage innovation and improvement processes leading to successful outcomes.

One of the reasons behind this struggle lies in the nature of the service process behind FM service provision and related innovation. According to the European standards, FM services deal with the integration of processes within an organisation to maintain and develop the services, which support and improve the effectiveness of its primary activities (CEN, 2006). Because of their supporting nature, FM services are characterised by a service process, which involves a heterogeneous range of stakeholders on both demand and supply side. In fact, each organisation requires a more or less formalised unit to take care of the FM services and ensure that its employees can carry out their core tasks and activities. Such a unit, the internal FM unit, carries the responsibilities of FM service provision, and when FM services are outsourced, it manages the relationships and outsourcing contracts with the external FM service provider(s). The internal FM unit thus plays a double role: (1) internal service provider in the eyes of the organisation and its employees and (2) customer in the eyes of the external service provider, with whom it negotiates the contracts at the basis of the service provision. Beside the internal FM unit, on the demand side of the FM service provision there are (1) the organisation as a whole, which orders and pays for the FM service provision, and (2) its employees, who eventually receive and take advantage of the FM service provision (see Figure B15.1).

This heterogeneity of the range of stakeholders, who are involved in the FM service process, complicates the management of innovation and improvement

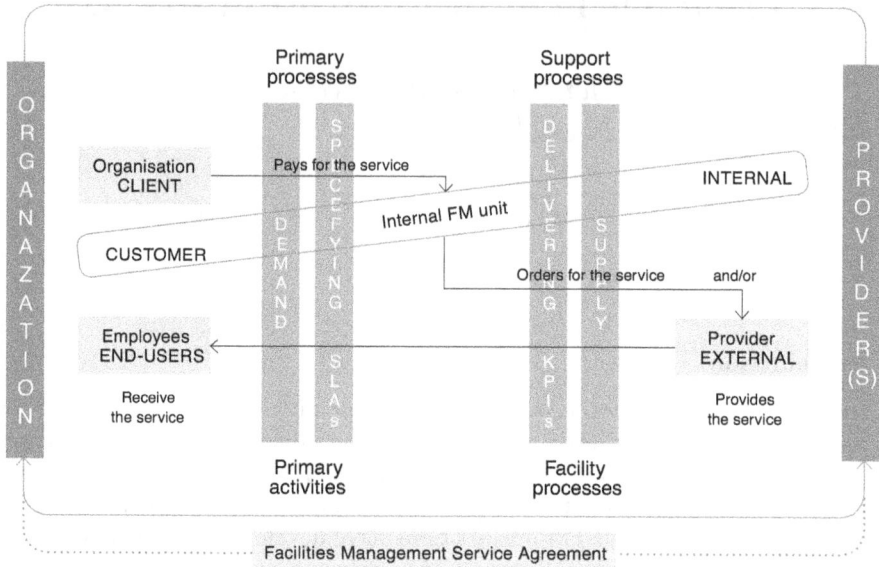

Figure B15.1 The FM service process

processes in FM. Every time a new process and/or service are introduced, the needs and expectations of all different stakeholders need to be taken into consideration. The complexity stands in the differences between these needs and expectations: clients, for example, typically focus on the overall ratio between FM costs and employee satisfaction of the facilities, while end users have much more individual needs and expectations, as they are the receivers of FM on a day-to-day basis. On the other hand, external providers care mostly about being able to efficiently provide satisfactory services to honour their contract with the internal FM unit. Consequently, every innovation and improvement process will have to contemporarily answer questions such as how will the process impact on the budget? What added value will it bring to the overall organisation? How will it affect individual end user satisfaction? How will it affect operational service provision and related costs?

Management of innovation in FM

So how can FM innovators ensure the success of innovation and improvement processes while dealing with such a heterogeneous group of stakeholders and their needs and expectations? All stakeholder needs and expectations should be taken into consideration to guide processes of innovation and improvement from idea generation to concept development and launch. To do so, the innovation developer(s), be it the internal FM unit or the external provider (or a combination of the two), should (1) clarify what different stakeholders need and what they would expect from the new process or service, (2) translate those needs and expectations into concrete goals and objectives and implement them throughout the innovation or improvement process and (3) demonstrate the fulfilment of needs and expectations to all stakeholders throughout the whole process by continuously communicating with different parties.

While this communication has traditionally been carried out after launch of the innovation, it would be recommended to start interacting with stakeholders earlier in the innovation or improvement process. This can be done by selecting specific actors to participate in the innovation or improvement process through dedicated tools, such as workshops and user surveys. However, such involvement should be carefully planned and managed. For instance, representatives of the client – i.e., the executive management – will not be interested in discussing operational matters, such as the type of food to be served in the canteen, but they should be closely involved in the strategic decision-making behind the shift from an independent canteen service provider to an integrated FM solution. On the contrary, end users have an individual view on FM that makes their contribution hard to process at the strategic level, but they can provide useful insight on day-to-day FM, being able to suggest, for example, a switch to environmentally friendlier time-limited taps or more efficient space management based on effective usage. Consequently, some tools for involvement are more appropriate than others depending on the stakeholders, whose contribution is sought, and on the nature of the decision-making to be carried out. On one hand, workshops will work best for an active discussion with top management on strategic issues. On the other hand, user surveys as well as mapping and profiling tools are most appropriate to (1) ask end users about the user satisfaction and hand-on experiences and to (2) observe their daily behaviour with respect to facilities and FM services.

Another way to support successful innovation and improvement processes is the development of dedicated structures, such as innovation platforms or innovation boards. A platform that is consistently managed and supported by mutual commitment from all parties can facilitate the development of innovation and improvement practices and processes over time, both within and beyond the boundaries of the platform itself. To do so, all potential stakeholders should be identified and actively involved from the very first steps of the development of the platform. In other words, representatives of all stakeholders, from clients, customers and end users to external providers, managing agents and even external contributors (e.g., researchers), should be invited to join the design of the platform so that they can contribute to its characteristics, values, vision and mission (see Figure B15.2). At the same time, diverse stakeholders often have different expectations and attitudes towards the platform which need to be identified and used to eventually create sub-groups of actors. Finally, tools and methods to facilitate interaction and contribution of stakeholders – such as, for instance, workshops or online competitions – should be selected, applied and modified depending on the circumstances and on the development of the platform over time.

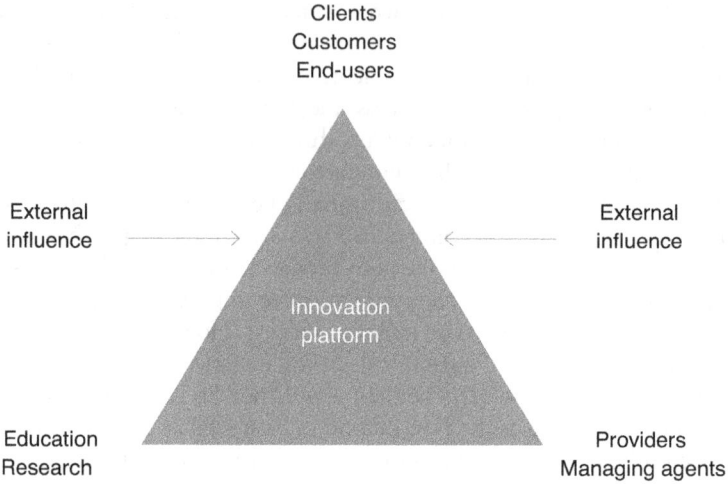

Figure B15.2 Example of innovation platform for FM

In summary, innovation and improvement processes are not just about creative ideas and project management. To raise awareness and successfully contribute to the core business of organisation, it is crucial that facility managers adopt partnership approaches that bridge between demand and supply when developing and implementing innovation and improvements. This can be done by involving selected stakeholders in the innovation and improvement processes based on how their needs and expectations fit with the type of decision-making to be carried out. Alternatively, dedicated structures can be built, such as an innovation platform, where representative stakeholders can meet, interact and share idea generation initiatives as well as concrete decision-making for implementation and launch. By doing so, potential tensions and conflicts due to the mismatch of needs and expectations between stakeholders can be overcome and even turned into drivers for innovation and improvement (see also Chapter B.14). The following is a case from the Netherlands of a successful implementation of an innovation platform.

The case of Essent

Essent, the largest energy company in the Netherlands, with ca. 3,600 full-time employees, provides private and business customers with gas, electricity and energy services. In 2011, Essent Facility Services (EFS, i.e., the internal unit of Essent responsible for facility services) outsourced most of the Essent facility service provision to one main managing agent, here fictitiously referred to as MA to ensure anonymity. After the outsourcing, MA was in charge of managing contracts with a total of six single providers of facility services and shared with EFS the responsibility of the strategic decision-making associated with facility service provision, improvement and innovation. Service quality, continuity and costs of facility services were all agreed on by EFS with the correspondent managing agents based on contracts that are grounded in specific service level agreements (SLAs). In addition, the managing agent designed individual contracts to manage the relationships with the single service providers.

With the aim of more easily dealing with the heterogeneity of its network of stakeholders, EFS launched an initiative, which they called the EFS Innovation Platform. The first sketch of such initiative was defined in September 2011 to include several stakeholders of EFS as well as some representatives from the research world. Since its ideation, the aim of the EFS Innovation Platform has been to engage with the diverse set of actors in its stakeholder network to stimulate innovation and improvement ideas, manage their development and implementation and monitor their outcomes for future development. In other words, the main goals of the platform were and still are (1) to stimulate cooperation within the facility service value chain, both on operational as well as on the strategic level and (2) stimulate all parties involved to come up with innovations that would increase the experience and satisfaction of Essent employees – i.e., the final recipients of the service provision.

Along with EFS, four different groups of actors were invited since its launch to participate in the platform (visualised in Figure B15.2): (1) the managing agent MA, (2) the six single outsourced service providers, (3) a variable number of representatives of academia and education specialised in facility services from two Dutch universities and (4) a variable number of outsiders called within the platform as *inspirators* – i.e., consultants and employees from facility service providers without a contract-related interest in the activities of EFS and/or MA. All participants work within the platform on a voluntary basis, which means they have no obligatory rules to participate next to the official contracts that providers and MA have with EFS.

The activities initiated by the innovation platform led to the development of a number of improvements. The results of the case study indicate that the interactions between the stakeholders were intensified both in width and depth as the stakeholders in the network came to know each other better, developed mutual trust across companies and learnt to reveal the needs and expectations of all parties and how to balance these through optimal solutions.

Literature guide

The chapter is mainly based on the following article:

Nardelli, Giulia: Innovation management or conflict resolution? How to take advantage of stakeholder interactions to drive FM innovation and improvement processes. *EuroFM Insight*, No. 34, September 2015.

The topic and the case are treated more in-depth in the following conference paper and scientific article:

Nardelli, Giulia and Marcel Broumels: Value co-creation: Ad hoc process or dynamic capability? A process case study. Paper presented at 7th International Process Symposium (PROS), Helona Resort, Kos, Greece, 2015.

Nardelli, Giulia and Marcel Broumels: Managing innovation processes through value co-creation: A process case from business-to-business service practice. *The International Journal of Innovation Management*, Vol. 22, No. 3, 2018.

Further information about Giulia Nardelli's research and publications can be found in Chapters B.14 and B.16.

Use of the method – assessed by Per Anker Jensen

As shown in the table that follows, the method is particularly useful in one of the five processes presented in Part A: process optimisation. The method can be used as an organisation and model for the management of innovation with the involvement of external parties. The method can benefit from being used in combination with the model in Chapter B.14 and the method in Chapter B.16.

Process	Phase							
Strategy development	A	B	C	D	E	F		
Organisational design	A	B	C	D	E	F		
Space planning	A	B	C	D	E	F	G	H
Building project	A	B	C	D	E	F	G	H
Process optimisation	A	B	C	D	E	F		

For process optimisation, the method can be used by managers of internal FM functions to organise innovation processes from the initial phase of evaluating current performance and improvement potential, and until the final phase of assessing the need for further optimisation (phases A–F).

B.16 Collaboration on value creation in FM innovation

Per Anker Jensen (based on work by Giulia Nardelli)

Introduction

No matter the industry, co-creation of value is a hot topic. Marketing specialists swear by it; manufacturers and service providers try to integrate it into their innovation practices; researchers study it. But what is value co-creation, and how does it apply to FM?

Value co-creation is grounded in the recognition that firms are no longer in full charge of deciding on the value to be offered to markets, but rather need to continuously cooperate with their customers, who become active collaborators in the creation of value. Value, in fact, is jointly created by supply and demand. The former offers the frame and resources for the co-creation of value, and the latter makes their needs and expectations explicit and shares their knowledge about how to satisfy them. In FM, value is created first and foremost by delivering and maintaining services that support the core business of organisations. According to the EN15221–1 definition, FM is 'the integration of processes within an organisation to maintain and develop (. . .) services which support and improve the effectiveness of its primary activities'. In other words, FM is expected to *create* value for the organisation it belongs to by at least delivering and maintaining services that support the core business. Moreover, FM can *add* value by contributing to the organisational performance it belongs to.

Chapter B.15 reflected on the challenges that the heterogeneity of the FM stakeholder poses for the management of innovation and stressed that all stakeholders' needs and expectations should be taken into consideration when managing innovation and improvement. Value co-creation is one way for providers to manage innovation and improvement processes together with their demand. In fact, the relationship between FM supply and demand is not straightforward. Internal FM units share supply – i.e. the service provision, with outsourced providers – while at the same time being their customer. On the demand side of FM service provision, moreover, we find the organisation to which the internal FM unit belongs and its employees, who, as end users, eventually receive and benefit from the services. This heterogeneity of stakeholders in the FM service process implies a variety of needs and expectations to take into consideration when innovating. Furthermore, it raises specific questions on the co-creation of value for FM innovators: *Who should we co-create value with? How should we organise for and manage value co-creation? Which tools and methods can we use to support a successful co-creation of value? How does value co-creation actually affect satisfactions of clients, customers and end users?*

The heterogeneity of the FM demand can actually be an advantage to FM innovators, as they have a large pool of stakeholders to co-create value with. FM innovators can, in fact, co-create value together with clients, customers and end users. The caveat is that it is not possible to co-create value with all stakeholders in the same way. On the contrary, what works for cooperating with top management does not yield the same results when applied to internal FM units or end users. Value is co-created with top management mostly when the purpose is support and legitimate strategic decision-making and planning behind new FM services and processes. End users, conversely, contribute the most when involved in more operational co-creation processes.

FM innovators can, therefore, organise for and manage value co-creation by distinguishing between needs and expectations of the different stakeholder groups, and planning cooperation activities accordingly. Table B16.1 synthesises some of

Table B16.1 Tools and methods for value co-creation

	User	*Resource*	*Co-creator*
Organisation as a whole/client	Ad hoc meetings	Transparency matrices and models	Regular and ad hoc meetings
		Workshops	Workshops
		Scenario analysis (with or without simulation IT)	
Internal FM unit/ customer	Workshops	Workshops	Workshops
		Shared training	Face-to-face meetings
		Team-building activities	ICT for information management and sharing
		IT for information management and sharing	Team-building activities
			Scenario analysis
Employees/ end users	User surveys	User surveys	Shared training
	User workgroups	Face-to-face interviews	Idea competition
	Workshops	Workshops	Workshops
		Idea competitions	
		Shared training	
		Team-building activities	

the support tools and methods that can be used to involve FM stakeholders in FM innovation. The tools and methods are classified in the table based on the role of the stakeholders and the type of stakeholder involvement.

The contribution of stakeholders as a resource is variable. Stakeholders can play the role of users through service testing and service support. When involved as users, stakeholders test the service and provide feedback based on their experience, which allows the FM innovators to improve their offering accordingly. When playing the role as a resource, customers are usually passive: providers investigate customers' opinions, needs and expectations through, for example, surveys or focus groups. Finally, stakeholders can be actively involved as co-creators and thus participate in various activities, from design to development of the new service. Customer-firm interactions in this type of involvement must be more intense and frequent, and the support mechanisms for such interactions are expensive, time consuming and technology intensive.

Overall, workshops are the preferred tool for most value co-creation activities, as they can be adapted in structure and functioning to specific contexts and needs. For example, workshops organised with outsourced providers and internal FM units can be used to involve the latter as co-creators – e.g. when training personnel for new service provision processes. Also, tools based on information and communication technology (ICT) can be used for value co-creation, mostly to support information sharing and management. In some cases (e.g., scenario analysis and transparency matrices), ICT-based tools are useful to facilitate communication between different stakeholders when facilitating value co-creation (e.g., between internal FM unit and top management of the client organisation).

The importance of partnership-like relationships with the outsourced providers, based on trust between individuals, is mirrored in the relevance of face-to-face meetings to support value co-creation between external providers and internal FM unit, top management and end users. However, top management needs to be involved in value co-creation through less 'requiring' tools, such as regularly organised and ad hoc meetings, facilitated through scenario analysis and transparency models. This seems to be due to the need to demonstrate the professionalism and value of FM services, along with the non-strategic focus that top management tends to attribute to FM services. Finally, end users can be involved in value co-creation through shared training sessions in which their knowledge, needs and expectations can be collected and shared, for instance, with the front-line employees of the external providers. End users are also sometimes involved as co-creators through idea competitions and workshops, which not only support the new service development but also increase awareness of FM within the organisation. Shared training and team-building activities facilitate direct involvement as resource and co-creators and allow for opening the innovation process while getting closer to the actual needs of the end users (see Figure B16.1).

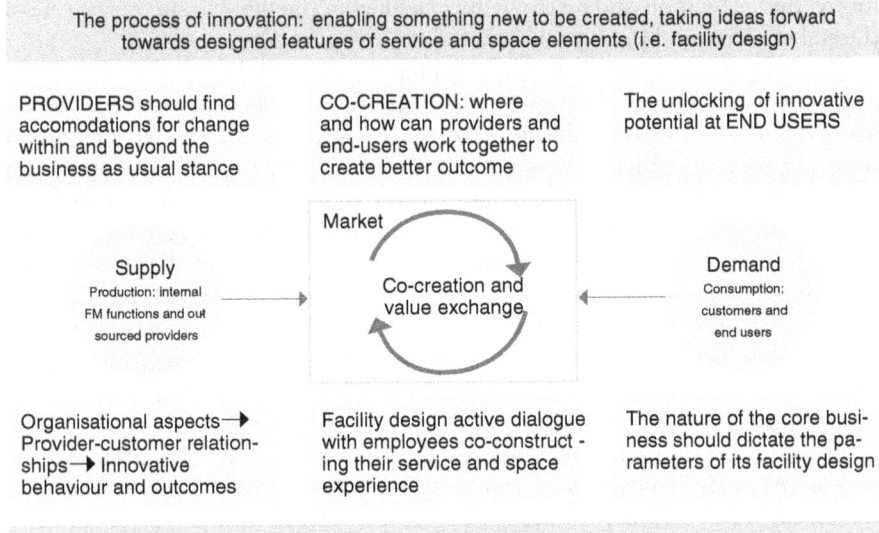

The process of innovation: enabling something new to be created, taking ideas forward towards designed features of service and space elements (i.e. facility design)

PROVIDERS should find accomodations for change within and beyond the business as usual stance

CO-CREATION: where and how can providers and end-users work together to create better outcome

The unlocking of innovative potential at END USERS

Market

Supply
Production: internal
FM functions and out
sourced providers

Co-creation and value exchange

Demand
Consumption:
customers and
end users

Organisational aspects→ Provider-customer relationships→ Innovative behaviour and outcomes

Facility design active dialogue with employees co-construct - ing their service and space experience

The nature of the core business should dictate the parameters of its facility design

Drivers and barriers of innovation: providers, end users and co-creation

Figure B16.1 The process of innovation and the co-creation of value with end users

Conclusion

In conclusion, value co-creation in FM is possible, and it can be implemented with all stakeholders. However, stakeholders' needs and expectations have to be matched and balanced for the value co-creation to be successful. FM service providers, who are in charge of FM innovation, should aim to co-create value with different stakeholders by paying attention to their different roles and to the possible degrees of involvement. Each group of stakeholders perceives and reacts to a different type of value depending on their needs and expectations. When the value offered by new FM services matches with the specific needs and expectations of a stakeholder group, their satisfaction is positively affected. Ultimately, it is the combination of value types that allows adding value to organisations – and value co-creation supports achieving such a combination as it brings providers closer to their stakeholder needs and expectations.

Literature guide

The chapter is mainly based on the following article:

Nardelli, Giulia: Value co-creation for FM in innovation: Is it possible, and if yes, how? *FM Update #1*, March 2017.

The topic is treated more in-depth in Giulia's PhD thesis and three conference papers:

Nardelli, Giulia: Innovation in services and stakeholder interactions: cases from facilities management. PhD Dissertation. Roskilde University. November 2014.

Nardelli, Giulia and Ada Scupola: User involvement and supporting tools in business-to-business service innovations: insights from facility management services. XXV International RESER Conference 2015. Copenhagen, 10–12 September 2015.

Nardelli, Giulia and Ada Scupola: Design tools for stakeholder involvement in FM innovations. *Proceedings of CIB FM Conference, Technical University of Denmark, 21–23 May 2014.*

Nardelli, Giulia and Ada Scupola: Involving users in complex service systems' innovation processes by means of ICT-based tools: The case of Facility Management services. SIG SVC 2013 Workshop: Workshop theme: Delivering and Managing Services in "Systems of Service Systems". ICIS 2013, Milan, 15–18 December 2013.

Further information about Giulia's research and publications can be found in Chapters B.14 and B.15.

Use of the method – assessed by Per Anker Jensen

As shown in the table that follows, the method is particularly useful in one of the five processes presented in Part A: process optimisation. The method can be used to organise innovative processes and choose the specific tools and forms of collaboration. The method can benefit from being used in combination with the model in Chapter B.14 and the method in Chapter B.15.

Process	Phase							
Strategy development	A	B	C	D	E	F		
Organisational design	A	B	C	D	E	F		
Space planning	A	B	C	D	E	F	G	H
Building project	A	B	C	D	E	F	G	H
Process optimisation	A	B	C	D	E	F		

For process optimisation, the method can be used by managers of internal FM functions to organise innovation processes from the initial phase of evaluating current performance and improvement potential, and until the final phase of assessing the need for further optimisation (phases A–F).

B.17 Personas as a basis for service innovation

Anne Vorre Hansen

Introduction

Within service innovation research, there are a variety of focus areas, among them innovation networks, employee-driven innovation and service innovation, in both the private and the public sectors. My PhD project was anchored in user-driven service innovation – i.e. with a specific focus on the role of the user, or customer, in the development of service. This relational focus has certain implications since the shared perception is that service, and herein the service experience, is framed by the relationship between customer and company. Thus a main distinction within service research is between service in singular – that is, service as a relational process and serv*ices* in plural – that is, the specific service offering given.

The PhD project was conducted in collaboration with the non-profit housing association Boligselskabet Sjælland and hence mainly based on empirical studies within the organisation. To ensure an ongoing resident focus in the development of the association's service offerings and communication activities I, in addition to the PhD thesis, developed three resident profiles.

Personas

The resident profiles are based on the persona method, which is a specific way to do target group analysis. A persona is a fictitious description of a user, customer or, as in this context, a resident. The personas method stems originally from IT development, but its application has spread to marketing, communication, concept development and innovation. Basically, personas support a shared understanding of your customers and that decisions in relation to development take as starting point, what the customers want and not what you and your colleagues think they want. Furthermore, personas makes it is easier to remember who the customers are, as well as their needs and motives.

In contrast to work with segments that mostly rely on quantitative data, personas are anchored in in-depth knowledge about the relevant target group, primarily based on qualitative data. The main aim of personas is, therefore, to obtain insights about the customer that are not solely related to demography. To exemplify, in the case of Boligselskabet Sjælland, the target groups were defined by the residents' expectations to and experience with the service relationship with the housing association – and not on gender, age, income or whether the residents were assigned from the municipality or not.

In the following sections, the three personas are presented. When reading the descriptions, it is important to remember that despite the fictitious element, all information is based on insights from resident interviews and statistics. Moreover, since the objective is to disseminate knowledge about the residents that can be integrated in development processes, no single persona can be found in a housing area but should rather be read as the sum of collected data. Thus in development, it is possible to take one or more personas as a point of departure to address either the whole resident base or to target communication to a specific group of residents.

Alongside the resident profiles presented in this context there are, both within Boligselskabet Sjælland and in the non-profit housing sector in general, resident groups with certain needs – e.g. socially marginalised and refugees. In the present personas, insight about these groups are not integrated since the main focus is on the majority of residents, whom employees and association refer to as 'the invisibles'. Nevertheless, the personas are generic and can be supplemented by either more profiles or by integrating specific needs and challenges into the existing profiles – as new knowledge is collected and emerges.

The engaged: Hans

Hans is 57 years old and lives in a terraced house with his wife Merete. They have lived in the housing department for 20 years, and their two children, both of whom have moved away from home, grew up in the department. Hans is a high school teacher in Roskilde. He originally came from Jutland, but moved to Copenhagen, where he studied. Here he met Merete, and when the children were toddlers, they lived in an apartment in the city. But Hans and Merete agreed to give the children new settings with more fresh air and safer school roads. At that time, they could not afford to buy a house, so they ended up getting one of the smaller apartments in the non-profit housing department. After seven years, they sought something bigger and got the opportunity to move to the current terraced house.

To Hans, a good home is a place surrounded by people who want you to feel good and where friendships can be built. Therefore, the feeling of home begins at the entrance of the housing department and not just at his own doorstep. It is important for Hans that neighbourliness is about pulling together, and he becomes annoyed when some of the residents back out of the community. Much of Hans' and Merete's social lives takes place with other residents from the housing department. When the children were young, Hans was not actively part of the resident democratic setup, but they enjoyed as a family taking part in social events. Later, when the children became teenagers, Hans entered the resident area committee, and he has now served on the board for eight years. In relation to the other residents and the culture of the department, Hans has been a focal point in the development of social cohesion. He knows most residents by name and has the reputation of being helpful and available, even at lean times of the day.   [1]

Hans would like to emphasise the positive elements of living in the non-profit housing sector. He argues that the diversity in the sector has given his children insight into other forms of life and thus a wider perspective than if they had grown up 'behind a hedge'. He knows the structure of the sector and understands that it can be difficult for non-profit housing associations to balance maintenance of the buildings and collaboration with residents. But at the same time, he is sceptical about the development of the housing association towards a more business-oriented organisation, because he is concerned about whether the management will take into account the resident democratic setup and whether there will be even less time for each department. The area committee has mentioned that the new development seems to make the area manager more stressed and the janitor less pleased with his work.

Expectations to the housing association

- That the housing association is 'meant' for residents – who pay the salaries of the employees and the operation of the administration
- That the housing association appears as a reliable expert in building maintenance and budgeting
- That the relationship between residents and employees is a collaboration based on equality
- That the residents' commitment is acknowledged and will be the basis for discussion and sparring
- That the employees of the housing association do not speak down to residents or act authoritatively
- That the employees of the housing association are aware that more residents have lived in the departments for many years and that they, therefore, know the area and the building very well

Quotation

I do not need to be trained in resident democracy; the housing association can just tell me what we are able to influence. Then I can deal with that, because we actually go into the area's committee, because we want to make the place we live in a nice spot.

Level of information

Hans would like to be updated on what is in general happening in the housing association, and he also keeps up with what is happening in the sector. As part of the area committee, he wants to have insight into decision-making processes, for example, in relation to the choice of subcontractors and time estimates. Hans knows that the area committee has limited influence on the budget, and therefore he would like to relate only to the part of the budget that the committee may in reality influence. Hans uses the home page of the housing association but does not participate in discussions on the housing association's Facebook page.

The social: Inger

Inger is 71 years old and lives in a 100 m² apartment in a housing department with 362 rentals. About 14 years ago, Inger and her husband Kurt moved from a service tenancy into an apartment in one of the other housing departments. When Kurt became a disability pensioner, they moved on to the apartment, where Inger has lived alone as a widow for three years. Inger is initially trained as a clerk but has had many different jobs throughout her life. Before she retired six years ago, she worked as home carer in Roskilde Municipality. Kurt's service tenancy was the setting for their common life and the growth of their three children, and Inger has many good memories from there. To her and Kurt, it was not important to own their own house, and Inger rather emphasises that they gained financial opportunities that they would not otherwise have had.

To Inger, a good home is a place where she can feel safe – both in her own apartment and in the housing department itself. Moreover, a good home is a cosy home 'with candlelight and where we are going to have candy!' Therefore, it is important for Inger to have room for guests and especially for children and grandchildren to come for visits. To Inger, neighbourliness means watching out for each other and the possibility to get help if the need arises. Inger has several good friends outside the housing department, and now, as a senior citizen, she participates in diverse events targeting her age group. But also, she has gained new friendships in the housing department, and when Kurt died, two neighbours, especially, supported her.

Living in the non-profit housing sector gives Inger, in her view, an economic latitude as a pensioner to have 'all the fun'. This means that Inger is able to afford going to the theatre, invite her grandchildren to the zoo and to travel on smaller trips. But how the non-profit housing sector is in general structured is not clear to Inger. Inger has participated in the annual budget meeting a couple of times, but in her experience, the atmosphere easily becomes a little unconstructive. Inger perceives the employees from the housing association to be nice when she calls, but she finds it difficult to get through by phone and often she experiences that no action is taken on her requests.

Expectations to the housing association

- That the employees of the housing association have the time and desire to speak and listen
- That the employees of the housing association know about the area and the buildings so that you do not have to explain how it looks each time you call

- That it is possible to reach the employees working in the housing areas and that they do their work properly so that the areas look nice
- That the housing association supports cohesion and social life in the housing departments
- That the housing association takes care of social problems in the department – and looks out for the weak residents

Quotation

> Earlier, you just called the caretaker and then only after an hour he arrived. Sometimes, there was even time for him to sit down and drink a cup of coffee. This cannot happen today.

Level of information

Inger is mostly interested in knowing what is happening in her own housing department – all information relative to the non-profit housing sector in general or the housing association as an organisation does not seem relevant to her. If she is to receive information, she would like to have a note in the staircase or a letter in the letterbox – but by far, the most appreciated form of communication is via personal contact!

The convenient: Camilla

Camilla is 29 years old and lives on the fourth floor in a three-room apartment with her boyfriend, Kristian, and their two children. Camilla is a receptionist, and she has just returned to work in an audit firm after maternity leave. Khristian is a locksmith with changing working hours. The children attend day care, and Camilla has just joined the parents' board. It was Camilla's mother who once wrote Camilla's name down to get an apartment in the housing association. Camilla grew up in a house, a little outside Roskilde, so she has lived in Roskilde all her life and finds it safe and nice for families with young children. Camilla and Kristian agree that as long as the rent does not rise too much, it is beneficial to be a tenant. But in the period in which they have lived in their apartment, the rent has been raised each year, and it concerns Camilla. She does not understand why it keeps increasing since she cannot see a corresponding increase in service offerings.

To Camilla, a good home is a place where you do not need to keep a straight face but can just be yourself. The apartment is the home itself – it's the family base where they can all recharge their batteries. Neighbourliness is to Camilla to say hello to each other and that there is a good atmosphere when you meet in the housing area. It is not important for Camilla to get to know her neighbours well, because she does not know how long they will live in the housing complex. Camilla gets a little annoyed by fixed rules and when the area committee acts as 'sticklers for the rules'. She would like a new playground, but because she is so busy with work, children and board work, she does not actively engage in change processes. However, she is pleased that the area committee arranges, for example, Shrovetide and an annual trip to an amusement park.

Camilla does not know the structure of the non-profit housing sector, and although she knows that she is living in the sector, it is not clear to her what it, in fact, implicates. However, she is convinced that it is somehow a safer way to be a tenant than in the private sector. Both she and Kristian are delighted with the right of disposal and the ability to put a personal touch on their apartment: they have painted the living room in a dusty blue, which Camilla thinks is beautiful. In the years that Camilla and Kristian have lived in the housing department, they have had virtually no contact with the housing association. But back in the day, Camilla became happy when she realised that being on the waiting list for a long time made it possible to get an apartment much faster than she and Kristian had hoped for.

Expectations to the housing association

- That the housing association is the landlord and therefore owns the buildings and determines a suitable rent
- That the housing association acts as a professional landlord with transparency in decision-making processes and with profitable agreements with subcontractors
- To be met by employees who provide good service with a smile on their lips
- That residents can quickly get in touch with a relevant employee and that employees will respond to enquiries in a short period of time
- That employees listen and are flexible in relation to individual needs

Quotation

I can just call as soon as there is something that needs to get fixed. I think that's very nice, and it makes me think it is really nice to be a tenant.

Level of information

Camilla is particularly interested in information related to her own home. This may be information about rent increase, water or electricity consumption – to Camilla, this is all related to the home, which is why she does not distinguish between the housing association or utility companies as a sender. Camilla prefers to receive information via e-mail, both because she can decide when to read it and also because the correspondence is hence documented. Although Camilla is active on Facebook and on various blogs, she does not associate social media with the housing association.

Literature guide

The chapter is based on the following report in Danish for Boligselskabet Sjælland:

Hansen, Anne Vorre: *Den engagerede, Den sociale og Den praktiske – en præsentation af tre beboerprofiler til brug i udvikling af relationen mellem beboere og Boligselskabet Sjælland*. Research report. Roskilde University. October 2015.

Further information about the PhD study can be found in the Anne's thesis:

Hansen, Anne Vorre: What's at stake? Critically exploring the concept of value co-creation in service research: Or an anthropologist going na(rra)tive in business administration. PhD thesis. Roskilde University. October 2016.

Other publications from Anne's PhD project include the following:

Hansen, Anne Vorre: "Back in the good old days . . ." – applying narratives to service research. Proceedings of XXIV. International RESER Conference, Helsinki, Finland, 11–13 September 2014.

Hansen, Anne Vorre: Context Matters. Exploring drivers and barriers for user oriented innovation within the public housing sector. ServDes 5th Service Design & Innovation Conference 2016, AAU, Copenhagen, May 2016.

Hansen, Anne Vorre: What stories unfold: Empirically grasping value co-creation. *European Business Review*, Vol. 29, No. 1, 2017, pp. 2–14.

Hansen, Anne Vorre: Narratives as driver for co-creating new stories of services. In F. Sørensen and F. Lapenta (eds.): *Research Methods in Service Innovation*. Location: Edward Elgar Publishing Ltd., 2017, pp. 76–94.

Hansen, Anne Vorre and Louise Li Langergaard: Democracy and non-profit housing: The tensions of residents' involvement in the Danish non-profit sector. *Housing Studies*, Vol. 32, No. 3, 2017.

Hansen, Anne Vorre and Luise Li Langergaard: Homeliness and self-determination: When conceptual discussions frame public service innovation processes. Conference on Public Service Innovation and the Delivery of Effective Public Services, Budapest, 15–16 October 2015.

Use of the personas – assessed by Per Anker Jensen

As shown in the table that follows, the persona method is particularly useful in two of the five processes presented in Part A: strategy development and process optimisation. Personas can be used as a tool to make development activities more goal oriented. The presented personas were developed in relation to the housing association, but the method can be used by other types of organisations.

Process	Phase							
Strategy development	A	B	C	D	E	F		
Organisational design	A	B	C	D	E	F		
Space planning	A	B	C	D	E	F	G	H
Building project	A	B	C	D	E	F	G	H
Process optimisation	A	B	C	D	E	F		

For strategy development, the method can be used by managers and staff, particularly on the strategic level, in FM functions to define strategy goals and develop strategy plans primarily in relation to new service offerings and communication (phases C–D).

For process optimisation, the method can be used by managers and staff in FM functions and their consultants in the initial phase to evaluate current performance and improvement potential and to identify and decide on changes (phases A–B).

B.18 Public-private partnerships and FM

Kristian Kristiansen

Over the last 15 years, public-private partnerships (PPP) have been the subject of many and intense discussions in Denmark, but very few projects have been realised. Nevertheless, there is still a political interest in PPP. But is PPP a good idea in general and in particular from a FM point of view? A research project at CFM has looked into this.

In Denmark, PPP is under the state's property enterprise and developer: the Danish Building and Property Agency. In 2012, this agency published a survey on lessons learnt from PPP projects in Denmark. Public clients believed that their PPP projects had a number of advantages. This, however, was without documentation or verification. The same agency followed up with another survey that looked into the assumed barriers for PPP: Given that PPP had so many benefits to offer and so few were realised, some kind of barriers had to be in play.

According to the Danish Building and Property Agency, 13 PPP projects were initiated in the period from 2005 to 2012. However, four of these were not financed by the private part and were not really PPPs but belong to a category that in Denmark is called public-private collaboration or integrated procurement. It is a characteristic of PPP that a private consortium – typically consisting of a contractor, a financing party and a property company – delivers a package for a facility consisting of financing, construction and management over typically 30 years, after which the facility is taken over by the public part. Following that delimitation, until 2013, one or two PPPs per year have been started in Denmark. (This has not changed much since then. In the period from 2012 to 2018, 12 new PPP projects were tendered.)

Good reasons for the small number of PPP projects

There are good reasons for the limited number of PPP projects in Denmark. A demand for private financing of public facilities is likely to be behind the many PPP projects in countries such as England, Australia and New Zealand. The policy of austerity has made it attractive to bring in new sources for financing. In Denmark, on the other hand, the Ministry of Finance has limited this possibility: In PPP projects, the public part has to deposit an amount corresponding to the construction costs unless specific permission is granted. Supporters of PPP have argued that this places an unfair disadvantage on PPP projects, but it can also be argued that under all circumstances, the public part has to pay for the interests, instalments and management related to the facility. Without the regulation, PPP would be like a credit card, allowing the public sector to run up huge debts.

High transaction costs are another reason for the few PPP projects. It is expensive both to tender PPP projects and to make bids: many issues have to be considered, many parties are involved and negotiations take a long time. Therefore, it has often been said that PPP projects should at least cost 200 million DKK (approximately 27 million Euro). In a Danish context, however, projects costing as little as 50 million Danish kroner (approximately 7 million Euro) have been considered.

A third reason is that it will in general be more expensive for a private part to finance than for a public body, and the private part will also need to make a profit. Both will place a huge burden on the reduction of management costs for the total cost of the PPP project to be competitive.

Lastly, PPP projects are inflexible for the public part. They tie up the procurer for a 30-year period to a contract with many obligations and a specific use of a facility.

So the question is whether the greatest 'barriers' for PPP projects in Denmark might be that few projects have the size needed and that the public part often prefers a certain flexibility for the use of their facilities. In countries where many PPP projects have been implemented, projects are often bundled in order to create the necessary volume for the PPP vehicle. In some cases, identical school buildings have been built in a number of municipalities. Not many Danish municipalities wish to build a school identical to a number of other schools and not designed to their own specific needs.

It should also be considered that the supply of projects needs to reach a certain volume before enough PPP consortia are on the market to ensure competition. In Denmark, one consortium has won the tenders several times, while some of the consortia that were interested in the beginning are now out of the market. Some of the later projects have been won by smaller companies that are not likely to take over more than one or a few projects. This raises the question of whether the Danish market for PPP projects is big enough.

Are the acclaimed advantages of PPP well founded?

In general, PPP projects will be more expensive because of the higher costs of financing, the higher transaction costs and the necessary profit. Supporters of PPP claim that the private parts are better at handling risks and that an integration of FM considerations into design and construction will trigger innovation and cost reductions to an extent that will make PPP projects economically advantageous.

Are these claims based on verified experiences? Unfortunately, no. A review of the literature on PPP clearly did not indicate only good results. Interviews conducted with participants in Danish PPP projects showed support to the ideas of integration of the design, construction and FM, and many reported a number of concrete problems related to the implementation of the good intentions. It was also underlined that it was practically impossible to win a tender through cost savings on management of the facility. A rule of thumb says that of the total costs of PPP projects, one-third is financing, one-third is construction and one-third is FM. Ten percent more efficient FM will only lead to a 3% reduction of total costs. And the public part will look not primarily at the costs of FM in the project, but at the project as a whole with its number of advantages and disadvantages compared to other projects.

A very reliable and thorough investigation of the results from ten years of PPP projects was conducted by the British government in 2012. It found that PPP had led to poor and expensive projects and a public debt of more than 60 billion British pounds.

PPP does not automatically result in an integrated construction process

It seems easier to understand why PPP projects have not flourished in Denmark than to understand why there still is such political pressure for more PPPs. An explanation for this might be that it seems so obvious that giving the responsibility for design, construction and FM to one actor – the PPP consortium – should lead to an integration of the various phases of the construction project. But as described in the research project, there are a number of factors that make project integration difficult.

Basically, a PPP is a legal agreement between a number of firms of the private part that form a consortium and another agreement between the private part and the public part. This does, of course, not mean a change of the physical organisation of the construction process – i.e. how the firms cooperate, how design and construction is organised, how the production process is completed or how FM is integrated. The research project found that in Denmark, the PPPs were run in much the same way as all other construction projects. Integrated teams or integrated designs that enable the inclusion of knowledge from all the participants are not a necessity in PPP projects. Nor is it a necessity of PPPs to form partnerships with or to integrate subcontractors in the process. And that a FM provider participates in the consortium does not automatically lead to a strong linking of FM considerations to the whole design and construction process.

Logically, four conditions need to be fulfilled in order to secure an integration of FM considerations in the construction process from the early planning phase to the end of the construction phase:

- Knowledge of buildings in use must exist. Some kind of systematic and verified knowledge accumulation in the form of, for instance, evaluations, key figures, data on costs, data on use, etc., must be available
- Knowledge of buildings in use must have a form that make it possible to be integrated with the many other kinds of knowledge that are needed throughout the construction process from the early start
- Knowledge of buildings in use must be in demand already up front in the planning process. The client, the architect, other consultants and other interested parties must want and wish to use knowledge on buildings in use

- Knowledge of buildings in use must be in demand throughout the whole construction process. All the participants, including the contractor and the subcontractors, must be interested in making us of the knowledge on buildings in use

Such an integrated construction process is difficult to establish. A formal arrangement, such as in PPPs, is far from being enough. In Britain, it was an important instrument in the various PPP programmes, such as PFI, Partnering for Schools and Procure 21 to motivate and coerce the participating companies to integrate the construction process: There was an awareness that PPP alone could not create the wanted integration of FM knowledge.

Prime contracting is another way of securing integration and has been taken in use by Defence Estates in Britain. The whole construction project is here in the hands of a so-called system integrator. That concept is known from similar delivery systems, such as delivery of rails, signals and trains in a complete transport system. The system integrator has a complete and well-documented supply chain. The supply chain is divided into clusters that can stand alone. In relation to the construction industry, the clusters could be, for instance, foundation, shell, building skin, installations and finishing. The system integrator has an extended responsibility that includes the early use phase in the form of commissioning or responsibility for management of the building in the first few years. Each cluster has a responsible firm in charge, and each cluster has an intimate cooperation between consultants, subcontractors end manufacturers. A cluster must compete with other similar clusters in order to remain in the supply chain.

In the research project, a comparison between PPP and other ways of arranging the construction process was made (see Table B18.1).

Table B18.1 Comparing PPP and other forms of organisation of the construction process

	Is there a mechanism that ensures knowledge about buildings in use?			
	Is it available?	*Can it be integrated with other forms of knowledge?*	*Is it in demand in the early phases?*	*Is it in demand throughout the whole construction phase?*
Construction systems	No	No	No	No
Strategic partnerships	No	No	No	No
Off-site production	No	Yes	Yes	Yes
On-site production	No	No	No	No
PPP	No	No	Possible	No
Prime contracting	Yes	Yes	Yes	Yes

Construction systems, such as systems for concrete panels, are developed separately and without concern for the total building or the building in use being a necessity.

Strategic partnerships are formal arrangements that do not necessarily lead to changes in the construction process.

Off-site production means that the factory delivers a finished building. All the phases of production take place under the same roof and have a single responsible party. This makes it obvious and easy to optimise the building as a whole. However, a mechanism for making the necessary knowledge available is missing – unless the client or customer is able to provide this.

On-site production is the traditional production form in the construction industry. Many parties are involved; the production process is divided into often unrelated phases; the use phase has an uncertain relationship to the production phase.

PPP – at least in the Danish cases – does not involve changes in the production process. It is possible that the existence of a consortium will mean that some considerations of the building in use will be made in relation to the bidding, but this is by no means crucial for forming a winning bid.

Prime contracting has a single responsible entity for the whole process and for fulfilling the demands of the client, also in the use phase. The supply chain is divided into clusters that has to be competitive, thus making integration of FM both possible and feasible.

PPP and FM

The research project concluded that PPP is not advantageous when seen from a FM point of view. PPP by itself does not lead to an improved integration of FM in planning, design and construction. Some interest in the use phase might be furthered by PPP, but as practised in Denmark, the necessary changes in the construction process have not taken place.

The interest in FM circles in Denmark regarding PPP is most likely related to a need for improving the understanding and acceptance of how important the use phase really is: It is in the use phase that the value for users is created, and huge amounts are spent for the management of the building. PPP is hardly a measure for fulfilling this need. If FM is to gain the importance that it should have, a new approach based on a deep understanding of how buildings create value for the user and evaluations of buildings in use could be a valuable measure. FM should make itself indispensable for an improved construction process.

Literature guide

The chapter is mainly based on the following summary report from the research project in Danish:

Kristiansen, Kristian: *OPP og indkøb af Facilities Management ydelser: Resumé rapport*. Centre for Facilities Management – Realdania Research, DTU Management Engineering. April 2013.

In addition, there are the following publications in English from the project:

Kristiansen, Kristian: To procure for better buildings: FM and public-private part-nerships in Denmark. Chapter 6.2 in Per Anker Jensen and Susanne Balslev Nielsen (eds.): *Facilities Management Research in the Nordic Countries: Past, Present and Future*. Centre for Facilities Management – Realdania Research, DTU Management Engineering, and Polyteknisk Forlag, January 2012.

Kristiansen, Kristian: Taking care of caretaking. ARCOM Annual Conference 2010, Leeds, September 2010.

Kristiansen, Kristian: PPP in Denmark: Are strategic partnerships between the public and private part a way forward? Proceedings from CIB TG72 Symposium: "Revamping PPPs", (ISBN: 978-962-8014-16-3), pages: 27–35, Hong Kong University Press, Hong Kong, 28 February 2009.

Use of the method – assessed by Per Anker Jensen

As shown in the table that follows, the method is mainly suitable for use in two of the five processes presented in Part A: strategy development and building project. The method can be used as an analyses tool and decision support in connection with strategies for and planning of building projects. The method can be used in combination with the model in Chapter B.21.

Process	Phase							
Strategy development	A	B	C	D	E	F		
Organisational design	A	B	C	D	E	F		
Space planning	A	B	C	D	E	F	G	H
Building project	A	B	C	D	E	F	G	H
Process optimisation	A	B	C	D	E	F		

For strategy development, the method can primarily be used by managers and staff on the strategic level in building client and FM organisations to evaluate current situation and define and develop strategy goals (phases B–D).

For building projects, the method can primarily be used by managers and staff on the strategic levels in building client and FM organisations in decision during the initiating phase of building projects concerning application of PPP or alternative ways to ensure FM considerations and integration of the supply chain and for strategic briefing and building briefing (phases A–C).

B.19 Collaboration with external providers

Kresten Storgaard

Introduction

Task execution in a modern enterprise typically requires an ever-increasing involvement of auxiliary functions and streamlining these can be a prerequisite for the company to deliver products at a competitive level. Some of these auxiliary features fall under Facilities Management (FM). There is thus a growing demand for efficiency enhancement and ensuring that they promote the company's main production. In some companies, it is through the professionalisation and efficiency of the internal management of FM tasks. In others, it is done through outsourcing these tasks to external companies. Collaboration with such external providers, therefore, becomes of increasing importance.

In Denmark, it is a political wish of the government that there is a continued outsourcing of tasks in the public sector through a competitive expulsion of the services in question. By competitive expulsion we mean that tasks are being offered on the private market. The goal is to obtain services that are both cheap and of satisfactory quality. Many of these services fall under FM. A large part of the efforts that are currently being made towards outsourcing in the public sector thus have a direct impact on the FM market and for the private companies that provide FM services.

This interest in outsourcing of tasks carried out so far by public organisations has largely set the agenda for the entire FM area. This applies both to the tasks that can be outsourced and what rules should be followed for the outsourcing process, taking into consideration existing (public employees) executives, tenderers and assignments, the decision-making process, grantors and end users, as well as the themes and topics that should be researched.

This has put focus in particular on the collaboration between public clients and private providers, and on the forms of procurement that are associated with the various forms of collaboration. Based on the policy efforts in recent years to increase collaboration between public FM clients and private providers, tender models have been prepared in accordance with the applicable procurement rules. However, these models do not pay particular attention to the challenges of establishing long-term collaboration, a type of collaboration which has the potential to strengthen the organisation as part of wider knowledge sharing, technological innovations and new, integrated services.

Especially the process of procurement is important if these challenges are to be addressed. For example, the offering determines the framework for the organisation and collaboration, and the client can prioritise between different skills for the collaboration, both own and the providers. We distinguished between five main forms of collaboration between public clients (mainly municipalities) and providers. The five main forms are as follows:

- Classical invitation to tender
- Service or operational partnerships
- Public-private company
- Design-build-operate (DBO)
- Public-private partnership (PPP)

It is worth noting that the vast majority of services provided by private FM providers to public clients are procured through classical tendering (80%–90% of all tenders). And it is worth noting that the public part of the overall FM market is actually limited. The vast majority of FM services are exchanged between private clients and private providers. But the conditions for public clients are quite different from the private clients. Public clients are bound to follow rules, including EU regulations, for procurement that ensure that all providers have the opportunity to bid, that all are treated equally and that the competition ensures a low price. Private clients do not have these restrictions. If you find trustworthy providers, you can continue to work together without putting the tasks out for tender. In practice, most private clients ensure that their providers perform the tasks at competitive prices – often this happens by contacting several potential providers – and closely follow price developments in the area.

Although the conditions for the public market and the private market are quite different, there are also areas where public and private clients can benefit from experiences in the other sector. The private sector can thus benefit from the experience gained in the public sector in terms of procurement and tender descriptions, benchmarking and quality assurance. And public clients can benefit from the experience of effective collaborations developed between private clients and providers, for example, in relation to long-term collaborations, where it has been found that there often is an opportunity to improve task executions for the benefit of both the client and the provider.

In the research project on long-term collaboration within FM, we, therefore, were interested in seeking knowledge about the experience of collaboration between companies and the influence this collaboration has on the production and the development of companies. As part of the project, a quantitative questionnaire survey was conducted among members of the Danish Facilities Management association (DFM), as well as four qualitative case studies. We present some key results from this in the next section.

The survey

The study distinguished between four types of collaboration:

- One-time tasks
- Continued tasks, without agreed upon time limits
- Continued tasks, with agreed upon time limits
- Operational partnerships

The four types of collaboration were chosen on the basis of interviews with key individuals. Operational partnership was described in the questionnaire as follows:

> Typically, agreement on common objectives for collaboration and task solution is included for optimisation of process and competence. Emphasis is placed on dialogue, openness and trust. The common objective of the operating assignment is based on the requirements specification with subsequent mutual clarification and clarification of expectations for operating results. Mission objectives may include user satisfaction, professional goals, method development, financial goals and service goals. Conflict resolution models typically enters into solution in dialogue at the level of conflict and use of a conflict staircase tool.

For the providers in the private market, the three former forms of collaboration each account for approximately 27% of the agreements, while operational partnerships account for 12%. For providers in the public market, continued tasks with agreed upon time limits account for the largest share, 39%, while continued tasks without agreed limit and one-time tasks each account for 22%. Here, too, operational partnerships are not widespread and account for 11% of providers' agreements. For clients, their purchases of FM services are distributed by 52% for continued tasks with agreed upon time limits, 22% for continued tasks without a time limit, 14% for one-time tasks and 8% to operational partnerships.

The study shows that there is a broad consensus on a pattern in which the collaboration through operational partnerships are generally assumed to have the greatest positive impact on task execution, while the one-off work generally is considered to have the least positive impact on this. But if you dive into the answers, it turns out that there are marked differences between providers and clients. And there is no agreement at all about what the buyer and seller consider they each get from the different forms of collaboration. Generally, providers consider that the form of collaboration has a far greater impact on the different factors than the buyers. And providers estimate that the operational partnerships have a more positive impact and one-off tasks have a more negative impact on job execution than assessed by the buyers.

Particularly, the difference between the providers' assessments of the importance of types of collaboration for the buyers, and the assessment the buyers make of their own benefits, may be of importance. The providers' assessment of the buyers' benefits of operational partnerships and the disadvantages of one-off tasks are not shared by the buyers. On the general level, buyers and providers agree on the pattern. However, when it comes to the concrete delivery of pros and cons, buyers are basically more negative in their assessment of what they themselves get from the operational partnerships.

One-third of the respondents are both buyers and sellers of FM services. Their answers more often resemble the answers from providers than the answers from buyers as regards their assessment of the positive impact of the operating partnerships and the negative impact of the one-time projects.

Case studies

In order to dive deeper below the results of the survey results, we conducted case studies of four companies. The cases were supplemented with an expert interview with Peder Stephensen, PS Experience, who was development manager in one of the leading Danish providers, ISS Denmark, during the period when this company evolved from basically providing cleaning services to becoming a leading provider of a broad range of facility services. The four companies were two buyers of FM, DATEA A/S and ATP Ejendomme A/S, and two providers, E&P Service A/S and Coor Service Management A/S. All case analyses can be read independently and contribute experiences that go beyond the conclusions of the summary analysis.

The two buyers of FM are responsible for real estate administration, including operations and maintenance, and for equity funds that have invested primarily in commercial properties. The two cases represent a market in which those responsible for operation and maintenance must pay close attention to the relationship between cost, quality and income in relation to the actual value of a property. This is an important difference to the public clients where there is no such focus on the relationship between quality in performance and own production, and no focus on fluctuations in the value of a property.

The two provider cases deal with companies that focus on delivering complex solutions – one with specialisation and focus on overall building maintenance and the other also focussing on integration of several types of FM tasks. Both providers and buyers see great benefits from long-term collaborations. But providers would like to see agreements that formalise a long-term period of time, preferably implemented in a true operational partnership. Buyers do not want to commit themselves to a formal agreement for a prolonged period of time but have experienced that they can get many of the benefits that can be found in long-term collaboration without the agreement on duration. Its more about the people who perform the tasks than its about contracts.

There are several benefits of long-term collaboration in itself. Efficiency rises because providers' staff get to know facilities, buildings and users, and thereby can organise and perform work more effectively and appropriately. Repeat effects are significant. Costs for supply and offerings are reduced. Contact between building and service users is improving, and customer satisfaction increases. The opportunity for knowledge sharing and collaboration has increased.

Summary

The results from the case studies highlight parts of the theoretical space that were presented in the project's initial report. The model shown in Table B19.1 expresses the expectation that there is a difference in the coherence between collaboration, knowledge and technology for standardised production – and for non-standardised production. Standardised production is coupled with a focus on contractually regulated forms of collaboration, a knowledge sharing based on formal education and explicit knowledge and a technology in which SLAs and KPIs are essential elements in the administration and control of task execution. In the non-standardised production, the form of collaboration is more regulated through relational capabilities, knowledge sharing is more about competencies and tacit knowledge, and technologically, the challenge is to focus

Table B19.1 Characteristics of standardised and non-standardised production

Performance	Standardised production	Non-standardised production
Form of collaboration	Contractually regulated	Relational capabilities • Among companies – and among persons • Flexible forms of collaboration
Knowledge sharing	Formal education Explicit knowledge	Competences, tacit knowledge • To be able to solve the task • Responsibility • Trust • To understand customers' and users' needs • Service mind-set • Knowledge sharing and experience building
Technology	SLA, KPI, Administration and control	BIM and visualisation, Embedded technology

on developing visual control tools such as BIM modelling and interaction with embedded technology.

The cases show that in the field of building operation, tasks and contracts are often perceived as a standardised production. However, it also appears that in cases where there is a recognised dependency on the use value of the service, providers are required to provide more than what is explicitly agreed – for example, in terms of accountability and initiative – when the situation requires it. Thus, it is a non-standardised production. The activities, such non-standardised production contain, are largely about providing services, which results in added value to the buyer. This added value is exactly the result of the competences and soft values that were highlighted in the cases.

This suggests that perhaps more attention should be paid to the differences between the two types of deliveries and markets, which are indicated in the difference between standardised and non-standardised production. And thus there is an awareness of the forms of collaboration, knowledge sharing and technology that relate to collaboration in the two markets. The cases also suggest that there is a fundamental difference between a private market where the quality of FM services has a direct effect on the client's business and where the client is aware of this relationship and actually acts in relation to it and parts of the public market.

Literature guide

The chapter is mainly based on reports from the projects in Danish, including the following report on the questionnaire survey and the report on the case studies:

Storgaard, Kresten: *Samarbejdsformer og Facilities Management – En internetbaseret surveyundersøgelse af leverandører og købere af FM ydelser.* Danish Building Research Institute. Aalborg University. SBi 2012:14, 2012.

Storgaard, Kresten og Freja Friis: *Længerevarende samarbejder inden for Facilities Management – Caseanalyse.* Danish Building Research Institute. Aalborg University. SBi 2012:15, 2012.

Publications in English from the project include the following initial report on theory and methods and a book chapter

Storgaard, Kresten, Jacob Norvig Larsen and Ib Steen Olsen: *Inter-firm collaboration in facilities management: Theory and concepts.* Danish Building Research Institute. SBi 2010:51, 2010.

Storgaard, Kresten and Jacob Norvig Larsen: Long-term buyer-supplier relations in facilities management. Chapter 6.4 in Per Anker Jensen and Susanne Balslev Nielsen (eds.): *Facilities Management Research in the Nordic Countries: Past, Present and Future.* Centre for Facilities Management – Realdania Research, DTU Management Engineering, and Polyteknisk Forlag, January 2012.

Use of the model – assessed by Per Anker Jensen

As shown in the table that follows, the model is particularly suitable for two of the five processes presented in Part A: organisational design and process optimisation. The model can be used as analysis tool and decision support in connection to sourcing and arranging collaborative relationships. The model can be used in combination with the method in Chapter B.9 and the model in Chapter B.26.

Process	Phase							
Strategy development	A	B	C	D	E	F		
Organisational design	A	B	C	D	E	F		
Space planning	A	B	C	D	E	F	G	H
Building project	A	B	C	D	E	F	G	H
Process optimisation	A	B	C	D	E	F		

For organisational design, the model can be used by managers and staff on the strategic level in analyses and decisions on sourcing and possibly when establishing long-term collaborations with external providers in the initial phases until the identification of need for new knowledge and competences (phases A–D).

For process optimisation, the model can be used by managers of internal FM functions to plan innovation processes from the initial phases of evaluating current performance and improvement potential to identifying and deciding on changes (phases A–B).

B.20 Maturity model for strategic partnerships

Jakob Brinkø Berg (born Jakob Berg Johansen)

Introduction

The road to strategic partnerships for the individual company in the construction industry in countries like Denmark may seem long, but that does not mean that you cannot plan and practice. To visualise how you, as a company, be it a building client or a business in the construction industry, can move from the typical way of working, we have created a maturity model. This maturity model was created by researchers at DTU based on literature studies and interviews with representatives of the participating companies in the societal partnership REBUS (Renovating Buildings Sustainably), which consists of housing companies and companies from the entire construction value chain, as well as knowledge institutions. The goal of the maturity model is to provide companies with a tool for analysing their own way of working and identifying the things they need to prepare for to participate in a strategic partnership.

The maturity model

The maturity model is shown in Figure B20.1. The five steps in the maturity model are five different ways to structure the relationship between the building client and the construction companies. The two axes, value and complexity, show that for each step, a more complex relationship is created, but at the same time, it is possible to create more value.

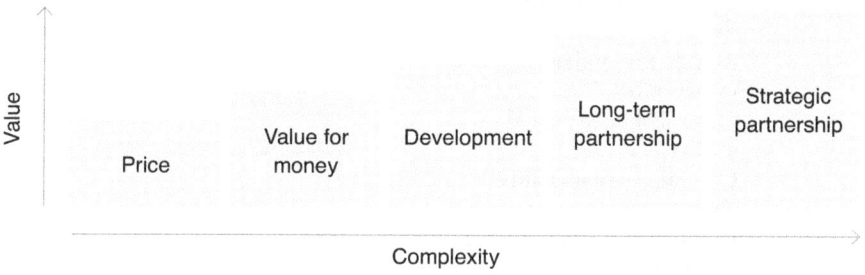

Figure B20.1 Strategic partnerships in construction maturity model (SPCMM)

To describe the two axes, eight key attributes (KAs) for complexity and ten KAs for value which companies using the development steps must be aware of, have been identified. These parameters can all be evaluated on a scale from 1 (very low) to 5 (very high) as shown in Table B20.1.

Table B20.1 Key attribute evaluation table for SPCMM

	Maturity level	Level 1	Level 2	Level 3	Level 4	Level 5
	Maturity requirement for level	Very low	Low	Moderate	High	Very high
Complexity	Complexity of tender bid evaluation					
	Number of qualitative tender bid evaluation criteria					
	Senior management involvement					
	Number of management communication levels					
	Goal alignment					
	Company value alignment					
	Quality of inter-organisational communication					
	Attention required to ascertain market prices of renovation*					
Value	Budget transparency					
	Transparent link between design changes and renovation cost					
	Transparent link between renovation cost and quality					
	Pre-renovation building assessment with input from multiple professions					
	Incentive structure with shared pain/gain					
	Risk sharing and holistic risk management					
	Control of price related to quality					
	Control of sustainability parameters					
	Development and testing of innovative solutions*					
	Dedicated renovation process facilitators*					
	*Specific KAs for project portfolios					

From the interviews conducted, the following actions have been identified, which can help underpin strategic partnerships:

- *Common goals:* It is possible for companies to formulate goals that are favourable for all participants in the strategic partnership.
- *Strong management commitment:* This enables employees at all levels to engage in the partnership.
- *Long-term commitment:* In order to make investments and create the necessary infrastructure for strategic partnerships, long-term commitment is essential.
- *Common positive incentive programmes:* To get a variety of stakeholders to strive towards common goals, a common positive incentive can be an essential tool for success.
- *Value and trust-based relational contracts:* The supply chain should be governed by contracts that strengthen strategic cooperation and thus based on shared values and build trust between stakeholders.
- *Shared data exchange platform:* One of the prerequisites for successful strategic partnership is the exchange of data and information.
- *Holistic risk management:* Managing risks across stakeholders can be a tool to minimise waste and costs because the individual stakeholders do not have to deal with risks individually.
- *Knowledge management:* In a supply chain, a conscious and systematic strategy for collecting and disseminating knowledge can help create cross-enterprise learning.
- *Systematic use of workshops:* To meet in person with a fixed agenda – but at the same time, there is also time for more unstructured interactions – is a very good tool at start-up, milestones and completion of a project.
- *Conflicts resolution model:* In an environment with many stakeholders, conflicts are almost inevitable, and a conflict resolution model, as well as a training programme for employees and leaders in resolving conflicts, can prevent costly delays and litigation due to conflicts.
- *Strategic partnership training:* Strategic partnerships require a unique set of qualifications to function successfully, and training programmes can be used to provide staff and managers with the necessary skills to handle strategic partnerships.

In Table B20.2, these actions are divided into strategic, tactical and operational levels.

Table B20.2 Actions that can support strategic partnerships

Strategic	Tactical	Operational
Common goals	Common positive incentive programmes	Systematic use of workshops
Strong management commitment	Common data exchange platform	Conflict resolution model
Long-term commitment	Holistic risk management	Strategic partnership training
Value and trust-based relational contracts	Knowledge management	

How to get started?

Although getting your business ready to engage in strategic partnerships requires many considerations, there are, nevertheless, some key skills that it may be a good idea to have in place to increase the chances of successful partnership.

- Procurement/contracts with several qualitative parameters that weigh values, competencies and trust
- Open books used to make qualified decisions. But remember that open books are not a good match for a traditional market contract relationship. A suggestion to practice open books in small scale is to make a development contract (SPCMM level 3)
- Whole-hearted management support and clarity about decision makers

These three points are not trivial for a company to execute and require in most cases education and focus. At the same time, however, research shows a lot of advantages when construction processes are developed to a higher level. Fewer conflicts, higher budget security, better user satisfaction and higher quality are just some of the benefits that can be derived from successful strategic partnerships. This does not mean that all companies or all building projects are suitable or able to enter into a strategic partnership, but for those who can, a strategic partnership is a powerful tool to have in the toolbox.

Additional information about REBUS

REBUS is an innovation consortium which has received funding from the Danish Innovation Fund. The work in REBUS is divided into a number of work packages that work in parallel with each other in order to create sustainable building renovation solutions. This chapter is based on the ongoing work in Work Package 1, which maps and develops new forms of collaboration and business models for the Danish construction industry. It is important to have good collaboration and business models for all parties that support the sustainable renovation in order for it to succeed. More information about REBUS can be found at www.rebus.nu.

Literature guide

The chapter is based on the following conference paper (please note that Jakob changed name from Jakob Berg Johansen to Jakob Brinkø Berg in August 2017):

Johansen, Jakob Berg, Per Anker Jensen and Christian Thuesen: Maturity model for strategic collaboration in sustainable building renovation 12. Paper presented at the 33rd Annual ARCOM Conference, Cambridge University, 4–6 September 2017.

The maturity model is being further developed, and there are plans to publish a final version in a scientific article.

Use of the model – assessed by Per Anker Jensen

As shown in the table that follows, the maturity model for strategic partnerships is particularly suitable for use in one of the five processes presented in part A: building project. The model can be used as an analysis tool and a basis to select partner companies. The maturity model has been developed in relation to strategic partnerships for building renovation projects, but it can also be used for other types of building projects, but it is aimed for partnerships concerning a portfolio of projects. Partnerships for FM in general are treated in Chapter B.19, and a maturity model for municipal FM organisations in relation to energy efficiency of buildings is included in Chapter B.11.

Process	Phase							
Strategy development	A	B	C	D	E	F		
Organisational design	A	B	C	D	E	F		
Space planning	A	B	C	D	E	F	G	H
Building project	A	B	C	D	E	F	G	H
Process optimisation	A	B	C	D	E	F		

For building projects, the model can be used by managers and staff on the strategic level in building client and FM functions, and in companies in the building supply chain before and during initiating building projects (phase A).

B.IV

Transfer of knowledge from FM to building projects

B.21 From FM to building

Per Anker Jensen

Introduction

Every facilities manager probably experiences now and then that there are parts or aspects of their buildings which do not function in an appropriate way. This can, particularly, in older buildings be caused by changes in the way the buildings are being used compared to how they initially were intended to be used. But often it is due to the professionals who planned the buildings not having the necessary insights to design the buildings in a way to made them well-functioning in relation to use and operation.

This is often made worse because of cost-cutting exercises during the building project involving changes, for instance, in materials and solutions which have detrimental consequences when the building is taken into operation. The common separation of the investment budget from the operational budget reinforces this. When the building project is managed solely with the capital investment in economic focus, there is a lack of incentive to make the building operation friendly, and when the investment budget is under pressure, it can easily lead to savings causing increased operational cost. In this chapter, I propose a model for mechanisms to transfer knowledge of FM to building projects to make final buildings more FM friendly.

The model

Within theory on knowledge management, a common distinction is made – like in economic theory and in technological development – between push and pull. In this context, this means that knowledge can either be pushed from FM to the people involved in building projects or knowledge can be pulled or demanded by the people involved in building projects from FM.

In addition, it is common within knowledge management to distinguish between personal knowledge and non-personal knowledge. Personal knowledge includes, among other things, so-called tacit or non-spoken knowledge, which is part of a person's competences more or less unconsciously. Non-personal knowledge, on the contrary, is available separate from persons, for instance, in written or digital form. One talks about such knowledge is codified. Thus, knowledge can be pushed from FM to building projects in terms of competences by directly involving persons with knowledge about FM or in a codified form.

Even though FM knowledge is being pushed by the use of competences and/or as codified knowledge, it is not sufficient for such knowledge to be utilised by the people involved in the building projects. It, first of all, requires that the designers and the building client are aware and realise the need to take considerations for FM into account or that the client uses his or her power to require that considerations for FM are integrated into the design process. Therefore, the prerequisite for knowledge being pulled into the building design is either awareness or use of power.

Based on these two forms of knowledge push – competences and codification – and the two forms of knowledge pull – awareness and power – I have developed a matrix with four types of mechanisms of knowledge transfer from FM to building projects, as shown on the left part of the model in Figure B21.1:

- Continuous briefing: A newer form of building briefing where users and facilities managers are involved in an ongoing dialogue with the designers in the development of a building. The difference between traditional briefing and continuous briefing is shown in Table B21.1.
- Detailed briefing: A brief document, which as far as possible specifies all requirements needed as a basis for the design of a building project.
- Project reviews: The client requires – and allocates money for – the review of documents from each design phase in relation to the fulfilment of considerations for FM, for instance, by involvement of specialist FM consultants.
- Regulation: Requirements are stated in the public regulation of building projects – for instance, generally for all buildings in the building codes or specifically for public buildings, for instance, that public clients are obliged to use life-cycle cost assessments.

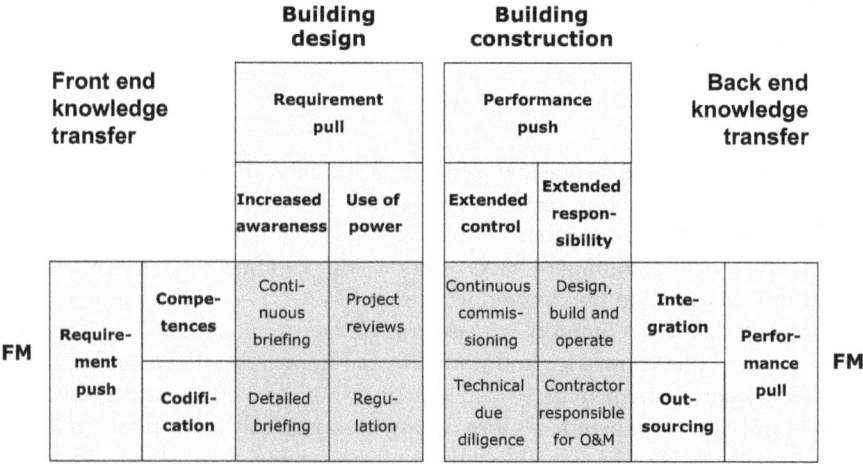

Figure B21.1 Model for knowledge transfer from FM to building projects

Table B21.1 Comparison between traditional and continuous briefing

Traditional briefing	Continuous briefing
Concerns new building/construction	Concerns all client/user needs in developing facilities
A definite phase at an initial stage	A continuous process with changing focus in different phases
An expert based information collection	A guided learning and dialogue process
Users mainly involved as data sources	Users actively involved as part of a corporate change process
The result is a brief, i.e. a requirement specification	The result is acceptance of solutions based on a brief

The four mechanisms, first of all, concern requirements for building projects in relation to integrating the consideration for FM into the design of buildings. Therefore, they are called front-end knowledge transfer in the model.

The right side of the model, in contrast, includes back-end knowledge transfer, where the focus is on validating the building performance. Here FM can pull knowledge either by integration of the coming FM operator with the parties involved in building construction or by outsourcing FM to a specific provider. Similarly, those involved in building construction can pull knowledge by increased control and extended responsibility. Based on these two forms of knowledge pull – integration and outsourcing – and the two forms of knowledge push – control and responsibility – I have developed a similar matrix with four types of mechanisms of back-end knowledge transfer from FM to building construction:

- Public-private partnership (PPP): The FM operator is an integrated part of the consortium with responsibility for design, construction, financing and operation of a building for a public user in long period, for instance 30 years, and therefore can set requirements for the designers and contractors for the building to perform as intended when put into operation.
- Contractor responsible for operation and maintenance (O&M): The contractor who is responsible for all construction as main contractor, for both design and construction as design and build contractor or for a specific part of the building work as trade contractor also has the responsibility of O&M of the whole building or the part they have delivered for a number of years, for instance 5, 10 or 15 years.
- Continuous commissioning: A newer form of commissioning where the FM function or a FM consultant are involved in a building project as early as possible to evaluate and validate the performance of the building's technical solutions – and, in particular, the building installations/services.
- Technical due diligence: A technical condition assessment of properties or a building in connection with a property transaction or the transfer of responsibility of building operations to a FM operator.

All these mechanisms of back-end knowledge transfer require, compared to the traditional handover of building projects, that increased focus is put on the performance of buildings being validated when the building is put into operation and occupied. Thereby, the involved parties in the building project get greater incentives to make sure that the building during both design and construction live up to the requirements defined for the building's performance.

Front-end knowledge transfer focuses, as mentioned, on the requirements of a building, while back-end transfer focuses on validating the building's performance. This difference in requirements and performance can be illustrated by energy consumption. The Danish building code set specific requirements in the energy frame of buildings in which the designers must document as fulfilled by energy calculations. Some clients define stricter requirements on the energy consumption in their building briefs, which the designers have to document in a similar way. However, there is no tradition for making the designers responsible for the actual energy consumption comparable to the calculated consumption when buildings are put into operation. This situation can be changed, for instance, with buildings developed as part of PPP. Here it is usually required that the PPP consortium fully or partially covers the cost of energy consumption. Therefore, we have had examples of PPP consortia holding the designers economically responsible for the calculated energy consumption being fulfilled.

Knowledge transfer from FM to building in a portfolio perspective

As shown in the model, continuous briefing and continuous commissioning are newer methods, which ensures that FM competences are involved in the whole duration of the building process to integrate the consideration for FM. For most FM functions, the primary focus is not on new building projects, but it is important that facilities managers have an ongoing focus on improving both existing and new buildings. This can happen as an ongoing collection of experiences with a form of continuous commissioning, where the performance of existing buildings is validated and optimised by regular intervals and where the experiences through continuous briefing are collected and formulated as updated requirements to buildings to be used in rebuilding, building improvements and new building projects.

A combination of the concept of continuous briefing and continuous commissioning in relation to the development of a company's property portfolio can be illustrated as shown in Figure B21.2.

The briefing process takes place during the use of existing buildings as an ongoing capturing of requirements based on experience and changing needs. When the need for a new building evolves, the briefing activity intensifies and has a peak around the start of the design phase but continues as a dialogue with designers during the design phase and to a certain degree with designers and contractors during construction. When the new building is occupied, briefing continues as an ongoing capturing of requirements in the extended portfolio.

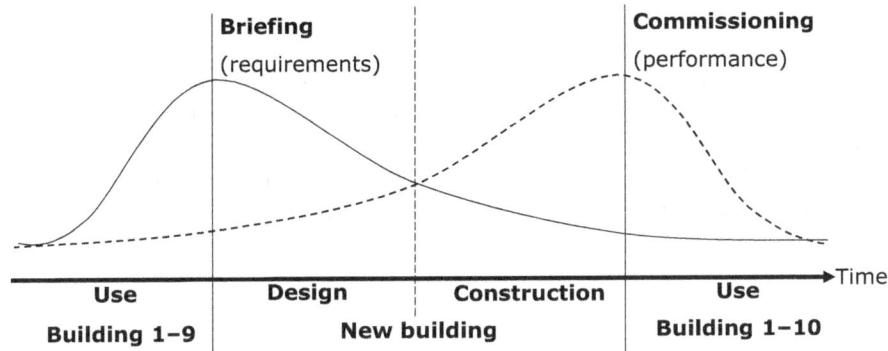

Figure B21.2 The pincer movement of FM on the building process

The commissioning process has a similar development but with an opposite intensity. During the use of existing buildings, it takes place as an ongoing optimisation of building performance and when a new building project starts, the commissioning process of ensuring and verifying the performance of the new building begins and intensifies during design and construction with a peak when the new building is occupied. When the initial building performance is verified, the commissioning continues as an ongoing optimisation of the extended portfolio.

Setting requirements for a building can be seen as a way to ensure that you get the right building, which is equivalent to effectiveness, while validating the performance can be seen as a way to ensure that you get the building right – i.e. a building that works right – which is equivalent to efficiency. The concepts of effectiveness and efficiency are treated further in Chapter B.27, where Value-Adding Management is seen as a form of management where both effectiveness and efficiency have a high priority.

Literature guide

The chapter is based on the following two scientific articles and a conference paper:

Jensen, Per Anker: Design integration of facilities management: A challenge for knowledge transfer. *Architectural Engineering and Design Management*, Vol. 5, 2009, pp. 124–135.

Jensen, Per Anker: Knowledge transfer from facilities management to building projects: Presentation of a model of transfer mechanisms. *Architectural Engineering and Design Management*, Vol. 8, No. 3, 2012, pp. 170–179.

Jensen, Per Anker, Torben Damgaard and Kristian Kristiansen: The role of facilities management in building projects. *Proceedings from Changing Role '09 Conference at Noordwijk aan Zee, The Netherlands, 6–9 October 2009.*

The concept of continuous briefing is treated more in-depth in the fowllowing book chapter and scientific article:

Jensen, Per Anker and Elsebet Frydendal Petersen: User involvement and the role of briefing. Chapter 8 in Stephen Emmitt, Mathijs Prins and Ad den Otter (eds.): *Architectural Management: International Research and Practice*. Wiley-Blackwell, 2009.

Jensen, Per Anker: Inclusive briefing and user involvement: Case study of a media centre in Denmark. *Architectural Engineering and Design Management*, Vol. 7, 2011, pp. 38–49.

Use of the tool

As shown in the table that follows, the tool is particularly useful in one of the five processes presented in part A: building project. The typology can be used as a management and analyses tool in the planning of building projects and possibly in combination with the methods and tools presented in the next three chapters.

Process	Phase							
Strategy development	A	B	C	D	E	F		
Organisational design	A	B	C	D	E	F		
Space planning	A	B	C	D	E	F	G	H
Building project	A	B	C	D	E	F	G	H
Process optimisation	A	B	C	D	E	F		

For building projects, the tool can be used by managers and staff on the strategic level in FM and client organisation, and their consultants in arranging the organisation, communication and form of collaboration in building projects from strategic briefing through all phases to evaluation (phases B–F).

B.22 Operational-oriented building process

Poul Henrik Due

Introduction

It is a well-known fact that knowledge about operation and experience from existing buildings in too few cases are used in the processes of new building, rebuilding and renovation. The result is that in many cases, solutions are chosen which are unnecessary expensive to run, doesn't fulfil user needs or solutions that must be rebuilt within the first year after handover. This issue was the start for a project in CFM that focussed on developing a tool for the screening of building projects in regard to future operation. The tool was called 'POKI', which is an abbreviation for Process, Organisation, Knowledge and tools and Information and communication. It is described in this chapter.

POKI

The tool introduces processes to ensure the implementation of operational issues in the building *Process*, the staffing of the *Organisation* and the behaviour of the stakeholders; the necessary *Knowledge and tools*, such as guides, checklists, software, etc.; and the necessary *Information and communication* activities. The tool is preferably meant for building clients and their consultants as they set the scene for a building project, but the tool includes recommendations for all stakeholders involved in building projects. Figure B22.1 illustrates the different phases and main elements in POKI.

From the very start of the *Process*, the client must set the necessary guideposts for an operational-oriented building process based on the vision, mission and values of the company. These guideposts form the basis of the demands in the design basis, which are mirrored in the basic design performed by the designers. The demands are maintained in the detailed design, the tendering process and the construction. Especially during construction, the focus is on time and economy of this phase alone, and the phase very often contains changes to the project due to financial cuts. Preparation for the future running of the facility is secured through quality assurance, commissioning, documentation and teaching of the future operational staff.

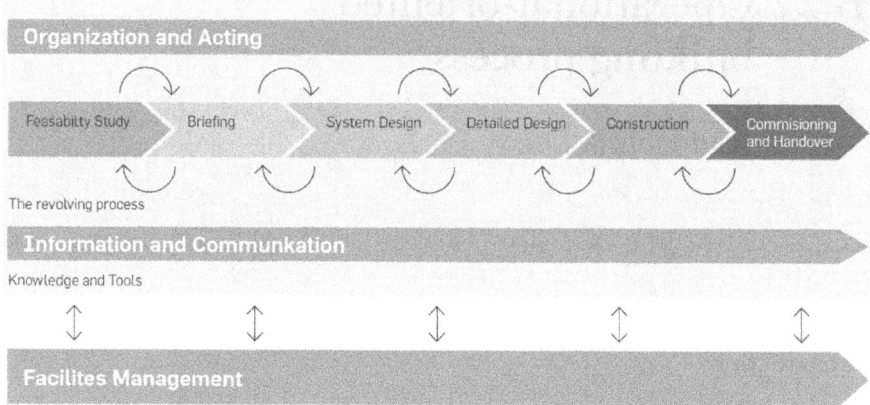

Figure B22.1 POKI with phases and main elements

The client's *Organisation* and manning of the project must mirror the wish for continuity in the project's focus on the maintenance of the operational guideposts set by the client in the basic design, thereby ensuring the visions for the facility are fulfilled. This, however, demands a strong engagement and a focussed steering by the client's top management. The management must appoint a committed leader who in cooperation with the consultants sets the right team – a team which must encompass one or more representatives for the future users and operational staff and if possible, the future head of the operational staff, a technical consultant and an operational consultant. Furthermore, management must ensure a 'linking pin' in the project from the start to the handover – a person with the right experience and knowledge to ensure that the intentions and information is transferred from one phase to another.

The third element of POKI is the necessary *Knowledge and tools*. To achieve an enhanced use of operational knowledge in the building industry, formal and valid knowledge must be generated by developing and implementing new digital tools tailored to assist the different stakeholders in the individual phases of the project from the basic design to the planning, construction, handover and running of the facility.

The final part of the POKI elements is necessary *Information and communication*. It is a natural part of a building process that there is an immense amount of information exchanged between the different parties and persons involved. The best possible information and communication effort must be based on a clear strategy and plan for a well-structured and prioritised communication between the parties in the project. Consequently, the client and the organisation must decide on a clear information and communication strategy and plan for the whole process.

Good ideas, but are they realistic?

POKI was presented for a group of experienced people from the building industry and one of the comments was that the tool is simply structured common sense, that the principles are general and that the management focus needed for the POKI to work is not realistic. So why develop POKI? Because several of the principles show their worth today in large, complicated building projects.

A case – The Maersk Tower

The Danish state's Building Agency is, in cooperation with the University of Copenhagen and the Maersk Foundation, creating leading research and teaching facilities for the Faculty of Health and Medical Sciences in Copenhagen. The Danish state is the client and will be responsible for the operation and maintenance of the building envelope. The university will lease and use the building, and be responsible for operation and maintenance of the technical installations, the indoor surfaces and the consumption of electricity, heating and water. Furthermore, the faculty must pay rent to the Building Agency. The consulting engineering company, Ramboll, and the architect firm, C.F. Moeller, won the task of designing the building. An illustration of the building is shown in Figure B22.2.

Figure B22.2 The Maersk Tower

During the project, both the CEO and the dean of the university have been very active in the process of preparing the competition brief, and substantial involvement of the future users as well as the FM staff of the university has taken place with the head of FM leading the efforts. The project has been organised with a steering committee, a builders' group, a commissioning group, a so-called Copenhagen University group and several user and research groups. The CEO and the dean were members of the steering committee and the Copenhagen University group. The users have been represented by a user coordinator, the head of FM and the Copenhagen University group with representatives for the users and the FM organisation. Furthermore, the FM organisation have been represented by the head of FM in the commissioning group.

The reason for the comprehensive involvement of the future users was, firstly, that the size, quality and complexity of the building should fit the rent that the university will be able to pay, and, secondly, that it is a complex building with comprehensive user demands. Consequently, there has been a strong focus on fulfilling the user demands and ensuring affordable life-cycle cost. A detailed description of the project organisation with precise areas of responsibility and profound information and communication plans were made. Furthermore, the client issued a set of standard requirements for the building based on many years of experience with this kind of building.

To summarise, there was the necessary management support, organisation and information and communication effort to ensure a user and operationally friendly building. This meant that most of the requirements suggested in POKI were met in this project.

It is realistic – and it pays off

It is possible – and maybe other clients are in need of user and operational friendly buildings. In Table B22.1, a small calculation illustrates that it pays off!

Table 22.1 An example: The construction of a 10,000 m² office building

Conditions	*Costs in million DKK*
Construction	150
Planning and design	15
FM during 30 years	300
Labour costs during 30 years	4,800
Investment in focus on FM	*Investment in million DKK*
+10% planning	+1,5
Savings due to a FM friendly building	*Savings in million DKK*
10% reduction in FM costs	−30
1% increased efficiency for the users	−48
Sum	**−78**

Additional information about the project

This chapter is based on the sub-project of a larger project titled Implementation of Operational Knowledge in Construction. The sub-project was a development project concerning best practices, and it involved workshops with a large group of facilities managers from a variety of public and private companies. The other sub-project was a research project at the University of Southern Denmark, and the results from this project are described in Chapter B.23.

Literature guide

The chapter is mainly based on the following book chapter:

Due, Poul Henrik and Peder Stephensen: POKI: A management tool for the implementation of FM know-how in construction projects. Chapter 7.4 in Per Anker Jensen and Susanne Balslev Nielsen (eds.): *Facilities Management Research in the Nordic Countries: Past, Present and Future*. Centre for Facilities Management – Realdania Research, DTU Management Engineering, and Polyteknisk Forlag, January 2012.

The full results of the sub-project are documented in the following two reports in Danish, including a best practice guide and a background report:

Due, Poul Henrik and Peder Stephensen: *Best Practice Guide – Implementering af driftsviden i byggeriet – DP2: Best Practice*. Centre for Facilities Management – Realdania Research, DTU Management Engineering, February 2011.
Due, Poul Henrik and Peder Stephensen: *Baggrundsrapport – Implementering af driftsviden i byggeriet – DP2: Best Practice*. Centre for Facilities Management – Realdania Research, DTU Management Engineering, February 2011.

Use of the model – assessed by Per Anker Jensen

As shown in the table that follows, the model is particularly suitable for use in two of the five processes presented in part A: building project and process optimisation. The model can be used as a guideline to plan the integration of operational knowledge and considerations in new building, rebuilding and renovation.

Process	Phase							
Strategy development	A	B	C	D	E	F		
Organisational design	A	B	C	D	E	F		
Space planning	A	B	C	D	E	F	G	H
Building project	A	B	C	D	E	F	G	H
Process optimisation	A	B	C	D	E	F		

For building projects, the model can mainly be used by managers and staff on the strategic and tactical levels in building client and FM functions from strategic briefing during pre-project through all project phases to evaluation in the post-project (phases B–H).

For process optimisation, the model, in particular, can be used in processes involving changes in buildings. It can be used by managers and staff on the strategic and tactical levels in FM functions in the initial phases of evaluating current performance and improvement potential, identifying, deciding and implementing changes and evaluating new FM performance (phases A–D).

B.23 Communities of practice as learning tool

Torben Damgaard

Introduction

Campus Kolding at the University of Southern Denmark has built a new university building close to the Port of Kolding, the river and city centre. The building provides the framework for research, teaching and the University of Southern Denmark (SDU). In connection with a research project that has followed the building process, it has become clear how the user role develops continuously in the process of the development of the building and how the changing role has a decisive influence on the process to be managed.

SDU in Kolding previously inhabited the town's old hospital. The university moved into the new building during the summer of 2014. One of the reasons that the university decided to construct a new building was that the old building was too small and some of the facilities in the building did not function optimally. The numerous small rooms and long corridors in the old building were not ideally designed for a modern university, and there was a shortage of space for the students. The previous university had the problem of being too close to the outside environment and was not open enough for the community to grow. At the same time, there were no optimal opportunities for expanding the university on the existing site.

In 2008, SDU organised an architectural competition in collaboration with the Danish Building and Property Agency, which is the building client. The competition was won by Henning Larsen Architects. Construction work started in 2012. Initially, the building will have a gross area of 13,600 m², which will house approximately 1,500 full-time students and 900 part-time students. It is further envisaged that the overall plan will eventually include a science park and stage in the university building. The cost of construction was DKK 280 million. The new university building is designed as a compact triangle and has six floors (see Figure B23.1).

At the heart of the building is an atrium courtyard that rotates with staggered floors down through the building as a large internal spatiality that spreads natural light from above. Each floor is designed with the intention of creating intersections between university teachers, researchers and students, while there are also areas for peace and contemplation. The individual floors are organised into offices and formal classrooms and team rooms, which are located on the periphery. When necessary, these can be opened up for more fluid study environment on the balconies where, among other things, there are various types of study workplaces and lounge areas located on the edge around the campus square. This means that all researchers and students have the opportunity to retire to the periphery or seek out social contact and interaction on the study balconies facing the atrium.

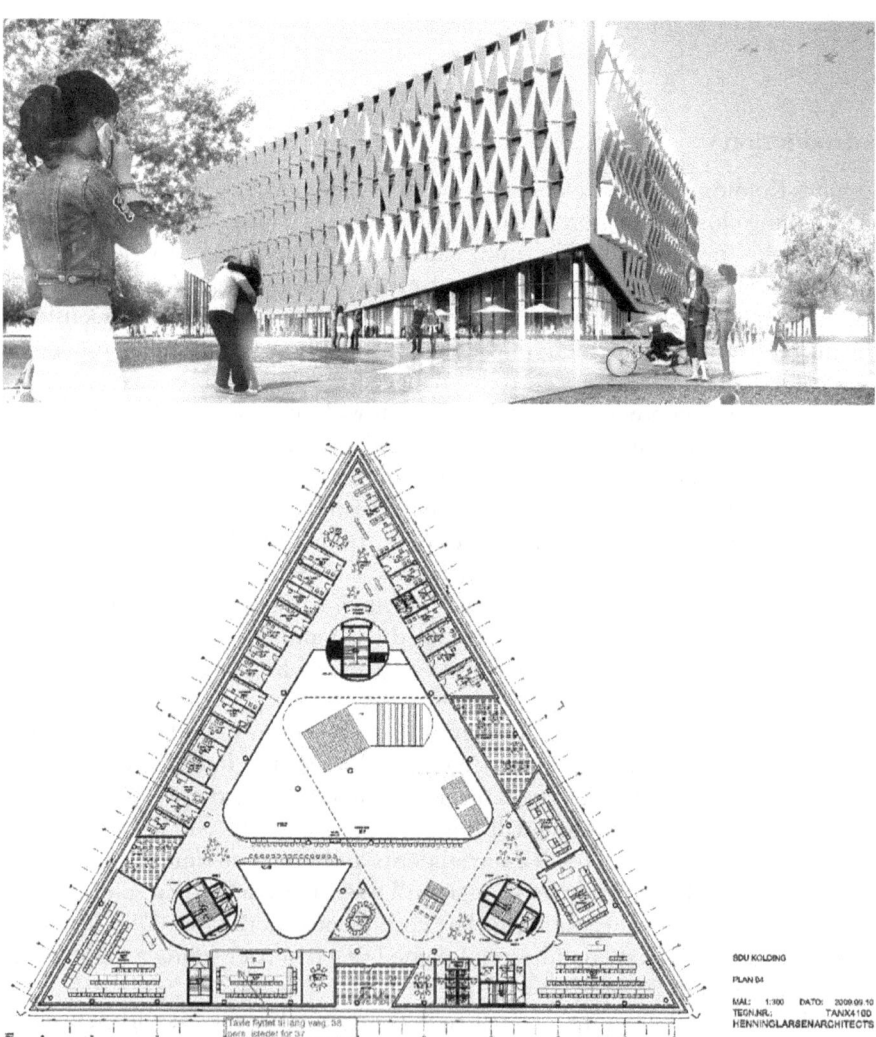

Figure B23.1 The Triangle building - photo and plan of the 4th floor (Printed with permission from Henning Larsen Architects A/S)

Research project

As part of the development of the construction project, researchers at SDU's Department of Entrepreneurship and Relationship Management on Kolding Campus have followed the project as part of a research project that has focussed on FM and user involvement. Researchers were, therefore, provided with the opportunity to participate right from the start and have subsequently followed and observed the development of the project through the entire process since 2008.

Some of the questions that have been consistent during the process have been as follows: *How may the knowledge of Facility Management be integrated into the various phases of construction through the entire development of the project? Who are the users, and who has sufficient knowledge to be able to represent the views of the users during the different phases?*

While clarifying who the users of the building are, it became clear in the process that several different users and stakeholders have been involved in the project. They have many different opinions, values and needs that characterise the wishes regarding what the future building will be able to offer, how it should be designed and the services that the university will be able to deliver in the future. The users have basically been the students, university teachers, researchers, management and members of the technical-administrative staff employed at the university, including the local IT and operations department. But since it is the Danish Building and Property Agency who is the building client together with SDU's building department, which is centrally located in Odense on a daily basis, they have also had an impact on the project, in addition to having different types of user requirements and wishes.

Similarly, by virtue of the visible role and the desired interaction with the surrounding community, the new university also has a number of external user groups and stakeholders in the local environment that are, among others, the business community, the surrounding institutions and the political system, etc. Finally, the group also consists of professional actors, such as architects, other consultants, the contractor, the subcontractors and the special portfolio stakeholders, as it is a major local showcase, and these could be described as special professional user stakeholders who have an influence on how the building may develop.

There are, therefore, numerous wishes and needs for the building. There are also many who seek to gain influence over what is going to be built. There are many users and user groups that have several and diverse requirements for how the task may be solved. This also means that FM and the operational task are very comprehensive and complex challenges to implement using the most optimum method. What is technically feasible to implement in relation to the requirements and specifications of the building legislation and the available finances must be measured against the specific requirements of the users.

User involvement in the project

In relation to the involvement of the users in the FM project, it was learnt that there are two parties in the building that are able to ensure that FM is put on the agenda and that FM issues are addressed: the users and representatives of the client's organisation. Both of these can play an important role in identifying the FM issues. The issues may be raised by the architect or another of the professional consultants but may also be raised by the users or users' adviser or by the client's consultant, who may also be the architect. In this case, where a university is being constructed, users may be represented by the client organisation (which also has an advisory function in relation to the building project and the selection of consultants) and the client's consultants may also be the architect.

It became clear that the users, client organisation, architect or client's consultant were able to take the initiative to address FM issues. But users cannot, as such, be expected to find solutions to the FM tasks. The users (defined in a broad sense as different users and user groups with various interests in the project) are very important actors in relation to addressing FM issues. On the other hand, the task of finding solutions to FM issues would require architects, engineers, consultants, advisers, contractors or actors who are involved in the operation and maintenance (including, e.g. some of the technical user groups and the university's technical organisation).

They would play various roles in the different phases of construction. For example, consultants and architects would be very important in the descriptive and idea-generating phases due to their professional understanding of the type of task to be solved. These actors may not be able to find the specific solutions, but their experience with the complexity of construction has provided them with a nuanced understanding of what and how FM can interact with other technical and formal solutions. The users play other important roles in these phases, as they are able to describe the cultures and values of the organisation.

Later on in the process, where the components of the building have to be solved and work together as a whole, follow-up research has shown that daily university users do not have the proper competences to solve the FM issues due to their complex nature. The university's FM organisation has entered into and assumed the role of user, particularly in the later phases of construction, as this requires more specialised knowledge. The role of user has, therefore, changed from being daily users and servicing staff to user experts or specialists who have in-depth knowledge of and experience with user problems and the more technical FM challenges.

At the same time, daily users in the form of management, university teachers, researchers, students and the technical/administrative staff groups have worked on strategies for what the building will be used for. The physical locations should be adapted to the changes that the organisation undergoes during the process. New requirements are being implemented for the integration of teaching and research, and this will affect the details of the physical framework. Thus, the users will also be essential in the later phases when decisions and solutions for specific FM issues are made.

FM development

Research has shown that the professional users, such as the client's consultant (architect), the construction company and the various consultants, *may be* significant representatives of different user perspectives and are therefore important as representatives of the users. They can act as bridge-builders for the users, as they often have a good understanding of the opinions and views of the user. Furthermore, the professional users have a good understanding of the entirety to be created and are often capable of considering the consequences for the overall system that the construction is made up of. This might indicate that all these actors should participate in the construction during the entire project, thus ensuring that FM issues are taken into account. It is not customary for all types of actors to participate in the construction in all phases of a project, which may also make good sense since it has to do with resource concerns.

Although FM staff members do not usually participate in all phases of the construction, the current case indicates that FM issues may still be well managed. The basis for the fact that FM aspects may still be well managed in construction is due to the fact that the architects, professional users and consultants are able to work holistically via education and experience. Working holistically is related to the way in which they practice their profession. It's not something they need to think about or plan, but it is something they are able to do by virtue of their professional experience.

There are, however, differences in quality that can be difficult to set targets for. The skilled professionals have the experience, knowledge, determination and motivation required, among other things, to manage FM professionally. It could be observed in the case building how various FM issues were continually identified and managed and how relevant persons were involved in finding solutions to the issues. However, involving all parties, actors and users at all times in order to solve the FM issues is not practical for carrying out a successful project. This is one of the reasons why faults that may be difficult to understand and explain are often found after the completion of complex constructions. There are thousands of decisions to be made during the building process and very few of these can be included explicitly in the planning. Most of them are made as professional decisions along the way and are based on professional expertise. This sometimes leads to some bad decision-making that subsequently results in constructional faults.

Problems have arisen in the Kolding Campus project, just as with all other buildings. An example of this is the lighting in the ceiling. Ever since the first drawings were made, the building has been constructed with a very large atrium room, where there are 24 metres to the ceiling. It is a room that winds its way up through the building via a triangular hole. Therefore, each floor partially overlaps the one below. It was only realised in the later stages that the light bulbs that need to be maintained in the ceiling can actually only be changed from the ground floor level. The architect's FM analyses identified a solution relatively late in the process that involved using a so-called spider lift to be able to change the bulbs. It was ensured by the structural engineer that the floor construction would be able to support the weight of the lift and that it would be able to come in and out, which are often solutions that could give rise to significant rebuilding costs if they are discovered after the completion of the building. The architect's involvement in the issue through a good holistic understanding of spatial architecture and tangible FM issues illustrates how FM can be managed without actual representation from FM.

A complex construction like the one in the case consists of so many thousands of decisions and solutions that they cannot be part of an explicit plan in any circumstances. Many of these issues are solved as they occur along the way because, among other things, there is a specific practice to ensure that they are managed. Against this background, there are a large amount of problems that are totally unnecessary to put forward explicitly. There is only one common solution which is based on the professional learning base, and this is used. During the process, we have seen, among other things, that FM issues are identified but not addressed, because they are expected to be solved at a later stage when other technical and functional decisions have been made, and more information and knowledge about how to solve them is available. Finally, there are also FM issues that are left entirely to be managed when it is decided which subcontractors will carry out the work. These issues are thus left entirely to these subcontractors to manage.

However, a number of issues regarding floating concrete floors, prototypes, ceilings, terrain lighting and finances were only specifically addressed when SDU's FM organisation entered the project. The architect was, therefore, met with opposition and cooperation at a time when they were involved in choosing contractors and subcontractors, as well as ensuring that the correct FM documents were available. In a number of cases, the FM organisation was also given the task of developing FM solutions together with the suppliers when they were selected.

User's role

Across the actors mentioned earlier and the presented examples, the follow-up research has helped to identify two key roles in the development of the building, which can help to ensure the management of FM activities. The first role that we identified was the 'role of facilitator', which is the one who manages the project. In the university building, this role has been carried out by the architect, which is often seen in other building processes. The facilitator does not necessarily find the solution to FM issues, but identifies them and may help to ensure that there are actors available to solve them.

Another important role that we identified was the '*role of constructively critical*', which is the one that questions the decisions being made. The role of constructively critical ensures that key issues are questioned along the way: actors who persistently ask questions about the unresolved or unmanaged issues. They question the decisions that have already been made or in cases where an argument exists for insufficient answers and solutions. This role of constructively critical may be carried out by the client's consultant in certain parts of the process, but in the case of the university building, it was carried out by the actor who is also the architect. Therefore, the architect is not suitable to carry out this role alone, as he would need to question himself. The architect, therefore, has a combined role of both the facilitator and constructively critical. This means that some roles in the building are equivocal and sometimes may even be ambiguous.

The architects may find themselves in situations where they cannot answer a question by themselves, for example. This could be an issue regarding where a specific room should be, where it should be located or how big it should be. This situation would require that another actor assumed the constructively critical role. The constructively critical role is not a role that is in opposition like an opponent that is trying to be an aggressive opponent or having other interests. Rather, it is a party that takes the role of providing a different angle to the problem or solution.

The study has also made it clear that everyday users are not always able to carry out the constructively critical role. This may, for example, be in situations where decisions concerning finances are required, or where complex technical insights or experiences with different solutions are required over time. This is specifically exemplified in the university building by an issue regarding the price of designing and maintaining a wall with plants. Another example has been in connection with problems regarding floating concrete. These are examples, where users have had insufficient knowledge of the nature and choice of options for the issue. In these situations, the interests of the users will be represented by other levels in the organisation or from the industry of potential suppliers.

This means that the roles in the construction process are constantly changing so that, for example, in some cases, the users are students and university employees, and in other cases, the local operations department and, again in other situations, SDU's operating organisation. The changing actors who are able to assume user roles, therefore, have an influence on the process that the facilitator will manage, as the roles in the process are constantly changing. This is illustrated in Table B23.1, where the resources and competences of the constructively critical actors and facilitators are described.

The model shows how FM issues are managed over time in a dynamic process where the constructively critical actors are in dialogue with the tasks of the facilitator. The table illustrates how the constructively critical actors change over time, which should be adapted to the tasks of the facilitator in the individual phases.

Table B23.1 How FM actors and tasks develop over time

	Feasibility studies	Programming	Design	Construction	Putting into service
Phase characteristics	Decision whether to build	Clarification of needs, specifications and requirements	Planning and design	Construction phase	Handover and operation
The constructively critical actors	Client and client organisation Stakeholders The political system	The users (students, university teachers, researchers, administrative and technical staff) Client consultant	Architect Engineer Advisers Consultants Client consultant	Client User organisation Contractor Suppliers Building authorities Client consultant	The users User organisation Authorities
Facilitator's tasks	Identifying Exploratory Investigative Ensure that everyone is heard	Ability to listen Be open Be curious Mediator in a value-based dialogue	Motivated Energetic Have the ability to end and maintain discussions	Manage Administrate Close discussions Decisive Controlling	Transfer to the future users Create visions See possibilities Create safety and security

Conclusion

As we have seen, FM issues continuously arise in the various phases of the con-struction, and it is therefore often indicated that FM staff should be involved in all phases. There are multiple reasons for the fact that FM issues do not necessar-ily require the presence of FM staff in all the phases. FM activities are sometimes solved systematically as part of professional and user practices. There are other times where the activities are not managed immediately, as there is a practice of finding a solution at a later date. Sometimes the issues are so extensive that they may require new and innovative solutions. This may result in the issues being left for the suppliers involved at a later date to offer a solution for how they may be solved. The solution of FM is not innovative and creative by nature, but it is crucial that FM issues are managed along the way, as major faults, costs and drawbacks would subsequently arise if the issues were neglected.

It has not been the purpose of the case study of SDU's building in Kolding to assess whether FM has been managed correctly. It is expected to show that there will also be FM issues that have not been managed or were only partially managed in this building. The observations in the case are naturally not alone but build on literature surveys and analyses of other buildings. Against this background, it has been possible to identify the two key roles: the facilitator and the constructively critical, which in the case can be identified as essential for whether FM issues are managed during the phases of the construction. Therefore, the answer to solving FM's management challenges in the building is not only to ensure that the FM managers or operating staffs participate in the development process from the out-set of the project. They may be present when it is functionally and organisation-ally relevant, which is shown by the fact that they usually participate in the later phases of the construction development.

Additional information on the project

This chapter is based on a sub-project of a larger project on *Implementation of Operational Knowledge in Construction*. The sub-project on the university build-ing in Kolding was a research project. The other sub-project was a development project on best practice, and results from this are presented in Chapter B.22.

Literature guide

The results of the sub-project are collected in a final report with the main text in Danish, but also include research papers published in English:

Damgaard, Torben and Pia Storvang: *Facilitering: Model for implementering af FM i byggeri*. Research report. University of Southern Denmark and Centre for Facili-ties Management – Realdania Research, DTU Management Engineering, 2013.

Publications from the sub-project in English include the following:

Damgaard, Torben and Anders Peder Hansen: Communities of practice as a learning challenge in construction projects: How FM knowledge can be integrated in the learning process. Chapter 7.3 in Per Anker Jensen and Susanne Balslev Nielsen (eds.): *Facilities Management Research in the Nordic Countries: Past, Present and Future.* Centre for Facilities Management – Realdania Research, DTU Management Engineering, and Polyteknisk Forlag, January 2012.

De Haas, Henning and Anders Peder Hansen: Facilities management in a service supply chain management perspective. *22th NOFOMA Conference, University of Southern Denmark, Kolding, 11–12 June 2010.*

Storvang, Pia and Ann H. Clarke: How to create a space for stakeholder involvement in construction. *Construction Management and Economics*, Vol. 32, No. 12, 2014, pp. 1166–1182.

Storvang, Pia, and Torben Damgaard: A framework to integrate and support the development of Facilities Management activities in construction. Clibyg Seminar. CPH, 2013.

Use of the model – assessed by Per Anker Jensen

As shown in the table that follows, the model is particularly suitable in two of the five processes presented in part A: building project and process optimisation. The model can be used as a guideline to plan the integration of consideration for FM in new building, rebuilding and renovation, possibly in combination with the models and methods in Chapters B.20, B.21, B.22 and B.24.

Process	Phase							
Strategy development	A	B	C	D	E	F		
Organisational design	A	B	C	D	E	F		
Space planning	A	B	C	D	E	F	G	H
Building project	A	B	C	D	E	F	G	H
Process optimisation	A	B	C	D	E	F		

For building project, the model mainly can be used by managers and staff on the strategic and tactical levels in building client and FM functions from strategic briefing during pre-project through all project phases to evaluation in post-project (phases B–G).

For process optimisation, the model mainly can be used by persons involved in building projects. It can be used by managers and staff at the strategic and tactical levels in FM functions in the initial phases to evaluate current performance and improvement potential and to identify, decide and implement changes (phases A–C).

B.24 Integrating operational knowledge in design

Helle Lohmann Rasmussen

Introduction and background

A large number of newly built facilities are not as operational friendly as one would expect. The international scientific literature describes a reliability gap between the calculated and actual energy consumption, where actual consumption is higher than expected. In addition to increased energy consumption, other aspects of operational performance have been recognised to be deficient. There may be a lack of functionality, poor indoor climate, operation and maintenance difficulties and poor cleaning possibilities. In other words, a gap between the expected and the actual performance of the buildings can occur.

In some cases, changes to technical installations or components are necessary to optimise operation of even brand-new buildings. Changes may consist of adjustment or replacement of parts of the technical installations, changes or addition of building parts or altered behaviour and use of the building. Some deficiencies cannot be fixed once the building is in operation, or the changes are too expensive to complete. Thus facilities managers and end users sometimes simply have to accept reduced operational friendliness.

Both researchers and practitioners argue that integrating operational knowledge in the design stage of new facilities increases operational friendliness. Early involvement of facilities managers is considered to support the design team and building client to make more precise predictions on energy consumption, environmental impact, operating costs and identify inadequacies at an early stage. Thus if it is foreseen that demands will not be met in the completed facility, adjustments to the design can be made. Some researchers even find that operating personnel should be assigned a central and leading role in construction projects and hereby suggest great changes in the way in which construction work is typically carried out.

The purpose of the study described in this chapter is humbler. The purpose is to identify concrete initiatives and methods to integrate facilities management knowledge in the early stages of construction projects already available to the industry and, furthermore, to investigate if these initiatives and methods are used by Danish building clients. The focus is placed on the building client, as prior research results stress that the building client is the most important actor in terms of operational viability. Furthermore, the focus of the study is the early planning and briefing phase, because the earlier knowledge is brought into the project, the greater the benefit it can have. It is expensive and difficult – if not impossible – to make changes to the design if deficiencies are discovered at a later phase.

The study and its results

The study included the following three steps.

1   A review of former research and existing literature to identify practical initiatives, methods or tools for ensuring integration of operational knowledge in early stages of construction projects. Besides scientific literature, the review included guidelines and handbooks for practitioners.
2   A case study of the facilities management organisation – Campus Service (CAS) – of the Technical University of Denmark (DTU). Document analysis and interviews were used to map out the use of the initiatives identified in step one. DTU CAS was selected as it both operates and maintains a large existing building stock, as well as manages a large number of new construction projects on campus.
3   A questionnaire survey among five newer public indoor swimming pools in Denmark, to expand the mapping of step two. Swimming pools are selected as they are particularly energy consuming and complex to operate, and, therefore, expectedly, there will be a high focus on operation and maintenance during planning and design.

From previous research and existing literature in the field, step one, 31 initiatives were identified. The DTU CAS case study added 11 initiatives, thus 42 initiatives, small and large, to be considered in early planning and briefing phases of new facilities. Clearly, there are already a large number of initiatives that can be implemented to ensure more operational friendly facilities. Looking at the five swimming pools and DTU CAS together, it is found that most of the initiatives are to some degree implemented in practice. However, some initiatives are only very limited or randomly used. It is clear from the study that not only must an initiative be implemented but also a dedicated effort to get the initiatives adapted and integrated with the individual building client or even in the individual project is required.

The building client is responsible for implementing relevant initiatives. However, the fact that the building client is rarely an individual eliminates the otherwise clear distribution of responsibility described in the literature. The building client in both the swimming pools and DTU CAS can be seen as consisting of three main parties, illustrated in Figure B24.1. There are several other internal actors in a construction project, but these three are the most important in terms of ensuring operational friendly facilities.

Top management is responsible for the building portfolio and decides and orders new buildings. The Building Client Division handles project management of

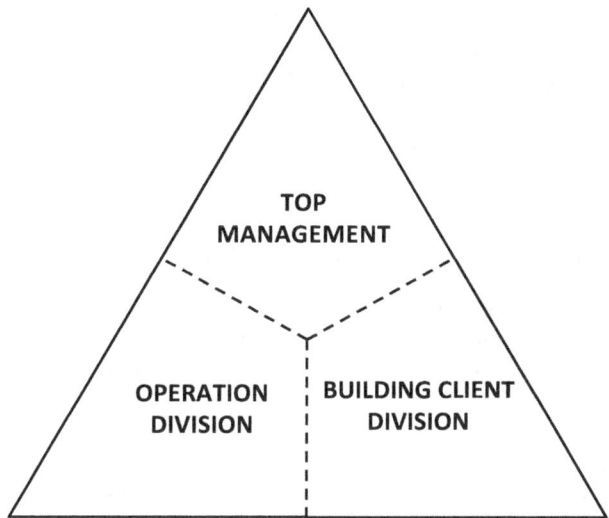

Figure B24.1 The three-partite building client

construction projects on behalf of top management. The Operation Division is responsible for operation and maintenance of existing and future facilities. In the swimming-pool cases, the Operation Division was de-centralised and located in the swimming-pool facilities, while the Building Client Division and top management were located centrally in the town hall. In DTU CAS, the three parties were co-located. The three-partite building client will be recognised in a large number of public and private organisations.

At DTU CAS, and probably in similar organisations, it is a common assumption that the Building Client Division is responsible for integrating operational knowledge in the design of new construction projects, as they are responsible for the projects. This study points out that it may be useful to revise that assumption and acknowledge that all three parties of the building client have important tasks in order to ensure that new facilities are constructed as operationally friendly as possible. When viewed in this manner, it is also clear that in a situation where one part is missing – for example, if the future operation division is not yet established – the other two parties must take care of the initiatives of the absent part, possibly by use of external competencies.

The three parties are not to be confused with the three commonly used terms for FM management levels: strategic, tactical and operational. For example, the list of initiatives that the Operation Division should consider includes initiatives at both the tactical and operational levels.

Among the 42 initiatives, 18 were categorised as 'well implemented' in DTU CAS at the time that the case study took place (2013). Thirteen of the 18 well-implemented initiatives were taken care of by the Building Client Division, 4 initiatives were handled by top management, while only 1 was taken care of by the Operation Division. It was not part of the research project to investigate why a lack of initiatives to ensure more operational friendly facilities initiated by the Operation Division were found, but possible reasons may be overshadowing focus on daily operations, lack of competencies or resources or a traditional assumption that the operational friendliness of new facilities is primarily the responsibility of the Building Client Division.

List of initiatives organised by responsibility

Table B24.1 presents a proposal for a revised list of the initially 31 + 11 initiatives that would be beneficial to consider in the early project phase of a new construction project. The initiatives are listed according to which of the three-partite

Table B24.1 Initiatives to integrate operational knowledge in design by the three-partite building client

Top Management should consider the following:	
1	Make a clear statement that operational friendly buildings are a high priority.
2	Establish a professional building client/construction management division.
3	Include both Operation and Building Client Divisions in the top management group.
4	Consider Facilities Management to be a strategic discipline.
5	Care for good relations between the Building Client Division and Operation Division.
6	Discuss life-cycle cost (LCC) with the Building Client Division in all projects.
7	Acknowledge that integration of operational knowledge – like user involvement – is time consuming.
8	Support public-private Partnership (PPP).
9	Contractor responsibility for operation and maintenance (Design-Build-Operate or Build-Operate).
10	Sustainability schemes, as LEED, BREEAM or DGNB.
Building Client Division should consider the following:	
11	Train project managers in how operational knowledge is expected to be integrated into every project, perhaps develop written guidelines.
12	Develop detailed building briefs.
13	Agree with the Operation Division about how and when they are making FM screenings of the project. If they are not, consider external FM screenings.
14	Use LCC as a decision tool.
15	Include a list of O&M demands in the building brief (get the list it from the Operation Division, see initiative 25).
16	Log deviations from the guidelines and standards for building projects (see initiative 24).
17	Establish a 'safety net' to secure considerations of comments and demands mentioned in the wrong phases and let the sender know what was decided.

18 Care for good relations between the design team, contractor team and Operations Division.

19 Evaluate the consequences of significant changes during the design phase or construction phase on operational friendliness. Inform the Operation Division about major changes and their consequences.

20 Facilitate involvement of O&M staff in workshops, meetings, mock-ups. Just like user involvement.

21 Implement incentive agreements for meeting specific requirements in contracts, such as indoor climate requirements.

22 Systematically study – and learn from – post occupancy evaluations (POE) of previous projects (see initiative 34).

Operation Division should consider the following:

---

23 Get acquainted with phases, actors, responsibilities, basic law and terminology of construction projects in general.

24 Develop (and maintain) general standards or design guidelines for new facilities in your organisation.

25 Develop a list of O&M demands to be included building briefs.

26 Specify requirements for the contractor's O&M manuals that fit your organisation, including the use of IT, and prepare to receive documentation from the project.

27 Specify requirements for O&M schedule and budget delivered by contractors or designers to ensure that they fit your organisation.

28 Review and comment on the design at different phases. Possibly use guidelines or checklists. Agree with the Building Client Division about which aspects of the design you review and comment on.

29 Demand/manage/participate in Building Commissioning. Bring your prioritised and measurable demands.

30 Perform a performance test. Request a coordinated test of the technical installations before handover. Preferably, participate in the tests.

31 Recommend extended supervision: Based on experience from previous projects, you can recommend extended supervision of selected parts of the project.

32 Plan who, when, how and why the named member of the O&M staff must be involved in the project. Allocate resources, construction projects are time consuming.

33 In cooperation with the project manager of the project, make a handover schedule, including training of operating personnel.

34 Prepare to evaluate the facility after occupancy (POE, see initiative 22).

35 Participate in evaluation or screening of design proposals in design competitions.

---

building clients should take care of the initiative. This is a simplification, as most initiatives require contribution from more than one part. Furthermore, the structure of the individual organisation can make the responsibility of an initiative best placed with a part other than proposed here as long as it is clear that the initiatives are spread on more than one part.

In the scientific field of knowledge transfer, two opposing approaches are dominant. One approach has a positivistic and technocratic point of departure and advocates that knowledge can be transferred, if codified, structured and shipped off by a sender and received at the appropriate time and in the appropriate context. Checklists, databases and design guides are rooted in this approach. The other approach has a social constructivist point of departure and has by some been labelled the behavioural approach. Those committed to the behavioural approach stress that knowledge cannot be transferred independently of the knowing person. They question the clear distinction of a sender and a receiver of knowledge and claim that in order for knowledge to transfer from one person to another, the parties must spend time together and develop a shared sense and value of the transfer. Face-to-face meetings, workshops and shared project offices are examples of initiatives rooted in the behavioural approach. The list of initiatives contains elements from both approaches based on the view that they complement each other and cannot stand alone.

The initial list of 42 initiatives was further developed. First, very similar initiatives were gathered to one initiative. Secondly, a number of initiatives relating to operation and maintenance demands to be included in the brief of a new project were gathered (No. 25) and are not further described here. Findings of an ongoing research project added a few more initiatives, such as sustainability certification and incentive agreements. The list contains both small and large initiatives, and even some that may be included in others (an example is life-cycle cost, which is included in sustainability certification) but can also be used alone. The list is helpful as a decision tool for the building client organisation in the very beginning of a new construction project. It assists the organisation in considering and deciding which initiatives are appropriate in their specific project, and which part is going to be responsible for implementing it. The initiatives listed here are all related to the early phase of a new construction project, as this was the focus of the study. However, the initiatives must be followed up continuously through the phases of the project, and additional initiatives may be important as the project proceeds.

Literature guide

The chapter is mainly based on the following scientific article:

Rasmussen, Helle Lohmann, Per Anker Jensen, Susanne Balslev Nielsen and Anders Højen Kristiansen: *Initiatives to integrate operational knowledge in design: A building client perspective.* Facilities, 2019.

In addition, the following two conference paper were also published from the project:

Helle Lohmann, Rasmussen, Per Anker Jensen and Jay Sterling Gregg: Transferring knowledge from building operation to design: A literature review. *Proceedings of IRWAS 2017 Conference in Salford, 11–12 September 2017.*

Helle Lohmann, Rasmussen and Per Anker Jensen: Tools and methods to establish a feed-forward loop from operation to design of large ships and buildings. Paper in Matthew Tucker (eds.): *Research Papers for EuroFM's 17th Research Symposium at EFMC2018, 5–8 June 2018 in Sofia*, Bulgaria. EuroFM.

Use of the tool – assessed by Helle Lohmann Rasmussen and Per Anker Jensen

As shown in the table that follows, the tool is particularly useful for two of the five processes presented in part A: building project and optimisation. The model can be used as a planning tool to prepare building and optimisation projects, possibly in combination with the methods presented in Chapters 21, 22 and 23.

Process	Phase							
Strategy development	A	B	C	D	E	F		
Organisational design	A	B	C	D	E	F		
Space planning	A	B	C	D	E	F	G	H
Building project	A	B	C	D	E	F	G	H
Process optimisation	A	B	C	D	E	F		

For building projects, the tool can be used by managers and staff on the strategic and tactical levels in building client and FM functions to prepare knowledge sharing from strategic briefing in pre-project through the whole project to evaluation in post-project (phases B–H)

For optimisation, the tool can mainly by used in processes that include changes in buildings. It can be used by managers and staff on the strategic and tactical levels in building client and FM functions to evaluate current performance and improvement potential, identify, decide and implement changes and evaluate new performance (phases A–D).

B.V

FM and added value

B.25 FM Value Map

Per Anker Jensen

Introduction

The main focus of Facilities Management (FM) has for a long time been on cost reductions of property operation and internal support services. This has, for instance, been achieved by extensive outsourcing of internal services to external service providers, which fused the development of a large and fast expanding market for facility services. However, we have within the last 15 years noticed a change towards the need for FM to create value or added value for the customers – independent of them being internal or external.

In a research project based on an explorative empirical study of FM best practice in the Nordic countries from 2005 to 2008, we investigated 36 cases from Denmark, Finland, Iceland, Norway and Finland. One of the general conclusions were that a change was occurring from FM primarily focussing on cost reduction towards also focussing on the creation of added value. This conclusion both concerned in-house FM organisations and external FM providers. The conclusion was underlined by the fact that the network of FM associations in the Nordic countries, NordicFM, in the same period, established a working group consisting of practitioners – with me as the only researcher – with the purpose being to 'highlight the added values to core business provided by Facilities Management'.

The difference between added-use value and cost reduction is illustrated in Figure B25.1. It shows the relative development over time of cost and use value of a service compared to a baseline with use value as specified in a service level agreement (SLA). The use value of the service can, for instance, be measured by a key performance indicator (KPI) with a minimum level of customer satisfaction. A cost reduction occurs if the cost/price of the service over time goes down without lowering the customer satisfaction below the minimum level. Contrarily, an increase in use value will occur if the customer satisfaction over time gets higher than the minimum level of customer satisfaction. This does not necessary involve a change in the SLA, but it means that added-use value is created.

In the following, I present some examples from the 36 cases on how some FM organisations have worked on the creation of added value. It should be stressed that this is based on a book from 2008 and therefore does not necessarily represent the current situation in these companies. Afterwards, I present the FM Value Map,

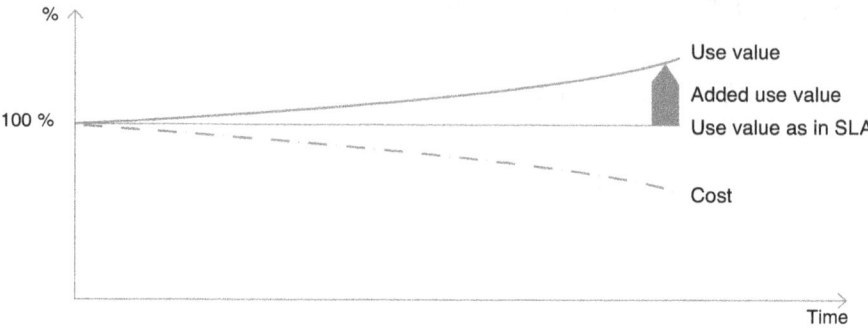

Figure B25.1 Added value and cost reductions

which was developed based on an analysis of these cases and participation in the mentioned working group in NordicFM.

Internal development of FM

The company Coloplast produces healthcare products and is a good example of a company that has put emphasis on internal development of FM. They want to be the best in their field to create growth and greater value for customers, staff and shareholders. The company has the opinion that the customers should have the greatest share of the value creation, but for the customers to obtain a satisfactory service and the right products, the company needs to have a highly motivated staff. Therefore, they emphasise the quality of the physical environment and the service and the offerings they bring to their staff on a daily basis. They have chosen to develop FM internally, and they have managed to document competitiveness, for instance, by using Lean principles, which are also used in the core business.

Another example is the Danish finance corporation Nykredit, which similarly to Coloplast has worked on developing the internal FM processes. They developed an equation called 'the user value ratio' as a tool for being able to measure and maximise the user value in all areas. This takes place through a holistic-oriented optimisation of each element of the user value ratio:

User value = Quality & Process/Price & Difficulties.

There are also examples with a focus on internal development of FM from public organisations. Both the Copenhagen Property in the municipality of the capital of Denmark and the Service Administration in the municipality of Malmoe in Sweden have as their objective to create added value to the customers who are the citizens in the municipalities – directly or indirectly represented by the administration of the municipalities. The Service Administration finds that the customer to acknowledge an added value must have a little extra besides just having his demands and expectations fulfilled. It does not need to be anything expensive. It is more the fact that the person who delivers the service shows consideration and plain humanity. It must be spontaneous and not based on instructions.

New collaborations with providers

Among FM providers, there is also an increasing focus on creating value for their customers. For instance, ISS on larger contracts offers to prepare a catalogue of improvement proposals annually, which they together with their customers decide on what to implement. At the same time, there is a strong development towards creating partnerships between clients and providers. PPP is a good example of a form of collaboration where life-cycle cost considerations over a building's lifetime comes into focus and thereby the building operation might get a higher priority.

Among municipalities, we have also seen a number of examples of operational partnerships. For instance, the municipality of Copenhagen has established partnerships with private consortia on operation and maintenance of all municipal buildings in different parts of the city for a period of years. The Danish pharmaceutical company Novo Nordisk has with success put the operation of office buildings out for tender based on functional requirements for a period of years. The common features of these forms of collaboration are that they are based on collaboration based on trust between the parties, better use of each other's competences, better possibility to adapt the activities to the companies' production capacity and capabilities and more value for the money for the customers.

In relation to energy, ESCO (Energy Service Companies) is a good example of a new concept for collaboration between a private company and a public organisation, where energy savings are achieved without the public part needs to pay directly except for a share of the guaranteed energy savings for a period of time. At the hospital administration in the Scania region in south-west Sweden, they learnt from experience that it is a good idea in ESCO contracts to involve the internal operational staff in a close collaboration with the private company so that the operation staff are upgraded and a long-term impact in ensured.

New areas of consulting

The Palace & Property Agency – now part of the Danish Building and Property Agency – see them self as the Danish state's facilities manager. Besides being owner and coordinator of external service provisions for many state organisations, they also offer their tenants consultancy on future office solutions, and they establish hubs in terms of buildings with shared offices with various common support services like reception and canteen for a number of state administrations, etc.

FM consulting is an increasing area of consultancy for consulting engineering company. At COWI – one of the largest Denmark-based consulting engineering companies – technical and environmental due diligence in connection to company mergers and property transactions has, for instance, become important as an area of consulting. Grontmij | Carl Bro – now part of the large Swedish-based consulting engineering company Sweco – has, among other things, developed a new IT system for asset management of technical infrastructure such as roads, water and sewage systems in municipalities.

FM value map

As part of the research project concerning FM best practice in the Nordic countries, the FM Value Map was established as a conceptual framework to understand and explain the different ways that FM can add value to a core business and possibly to the surroundings. The FM Value Map is shown in Figure B25.2. It is structured with cause-effect relationships with inspiration from the method 'Strategic Mapping' from the Balanced Scorecard developed by Kaplan and Norton.

The FM Value Map shows from the bottom that FM use certain resources that work as input to some processes, leading to a number of provisions as outputs. These provisions can as outcome have various impacts on both a core business and the surroundings. The impacts can be of benefit to various stakeholders.

On a more detailed level, the resources of FM are subdivided into facilities and activities with facilities consisting of real estate and technology, whereas activities consist of manpower and know-how. The processes in FM are subdivided according to the quality circle of PDCA (Plan, Do, Check and Act, see also Chapter B.28) in planning, coordination, controlling and improving. The provisions of FM are subdivided in basic products and additional offerings. The basic products consist of space and services, whereas additional offerings consist of development and relationships.

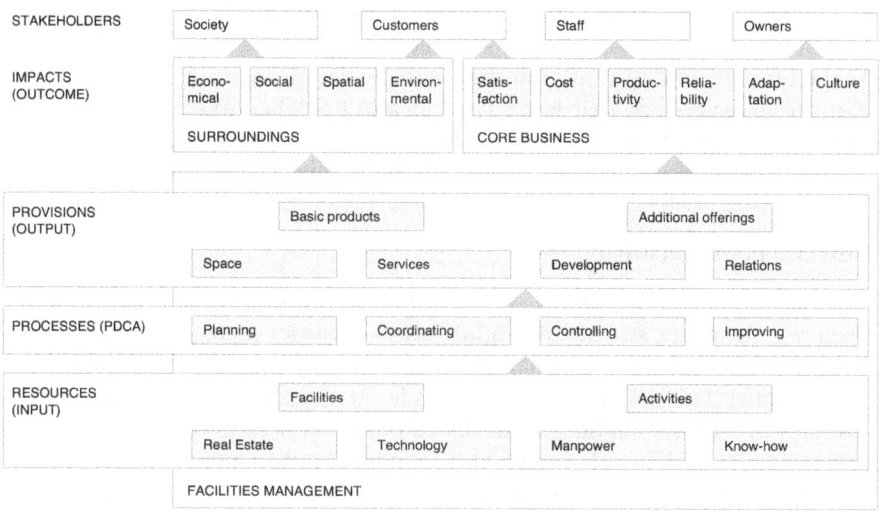

Figure B25.2 FM Value Map

The FM Value Map is based on the distinction between FM and core business, which is an essential part of FM theory. The FM Value Map further distinguishes between the effects of FM on core business and on the surroundings. It shows that the provisions from FM can create added value by contributing to effects on core business in terms of satisfaction, cost, productivity reliability, adaptation and culture. For the surroundings, FM provisions can similarly contribute with economic, social, spatial and environmental impacts. The stakeholders, who can benefit from these impacts and thereby obtain the created added value, are divided into owners, staff, customers and society.

The FM Value Map shown in Figure B25.2 is generic and can be used as an analysis tool to identify and illustrate the different ways that FM can create added value. This can, for instance, happen by pinpointing the elements in a FM provision that contributes to create added value with specific impacts with benefits for specific stakeholders. In this way, case-specific value maps can be produced. This is particularly relevant when a change in FM provision has occurred or is being planned.

Conclusion

The development from the financial crisis started in 2008 meant that the focus within FM again for a period became targeted dominantly towards cost reductions. This has since changed back again, and there is for me no doubt that the FM profession in the future will be forced to improve its competences to create added value. This is a necessity if FM is going to be an important discipline that is capable of obtaining the attention of the top managers in corporations and other decision makers and attract demanding and highly qualified young staff to the field.

This change has enormous consequences for the need for knowledge within FM. Cost reductions can be achieved by use of general management principles and methods used in other industries; however, added value can only be achieved based on specific knowledge and methods related to this particular field of practice. Furthermore, such a development cannot solely be based on knowledge from practical experience. Creation of new research-based knowledge and development of a body of knowledge specific to the FM profession is necessary.

Until now, the FM providers have been able to expand into new market areas due to the still increasing outsourcing without being forced to make radical innovations. This will change along with the market becoming more mature and saturated. The trend towards increasing focus on sustainability and corporate social responsibility also requires that FM develops new knowledge and competences to be able to make relevant contributions to the development and value creation in companies.

It is also important that FM professionals are capable of explaining the benefits to their clients, customers and end users of the services they provide. The FM Value Map presented in this chapter is a conceptual framework and an example of a tool which can be used to support the dialogue between the supply-and-demand sides of FM. The intention is that the FM Value Map can help to give legitimacy to the FM discipline among decision makers outside of but with relations to FM. However, the FM Value Map is also seen as a support to the internal development of the FM profession to become more conscious about the impacts of their activities and provisions.

Literature guide

The chapter is mainly based on the following book – published in English and Danish – from the project on FM Best Practice in the Nordic Countries:

Jensen, Per Anker, Kjeld Nielsen and Susanne Balslev Nielsen: *Facilities management best practice in the Nordic countries: 36 cases.* Centre for Facilities Management – Realdania Research, DTU Management Engineering, 2008.

The FM Value Map is also presented in the following scientific article:

Jensen, Per Anker: The facilities management value map: A conceptual framework. *Facilities*, Vol. 28, No. 3/4, 2010, pp. 175–188.

See also the following chapters in this part of the book for further information on CFM's research and publications on the added value of FM.

Use of the method

As shown next, the FM Value Map is particularly useful in one of the five processes presented in part A: Process optimisation. The method can be used as an analysis tool, possibly in combination the models in the following four chapters.

Process	Phase							
Strategy development	A	B	C	D	E	F		
Organisational design	A	B	C	D	E	F		
Space planning	A	B	C	D	E	F	G	H
Building project	A	B	C	D	E	F	G	H
Process optimisation	A	B	C	D	E	F		

For process optimisation, the method can be used by managers and staff in FM functions and their consultants in all phases from evaluation of current performance and improvement potential to evaluation of new performance and assessment of need for further optimisation, and as a basis for dialogue with internal and external stakeholders (phases A–F).

B.26 Adaptation between FM product and process

Akarapong Katchamart

Introduction

To be the best at adding value to the core business, the 'being strategic' manifesto has been a vibe in seminal FM works and practitioner's discussions during the last decade. But is 'being strategic' always the right answer? Is it always bad (as if) to just be cost driven? Is FM truly critical to the business? This chapter argues that the answers to these questions are depending on the context and the type of FM product that we are talking about. The chapter presents a quick and dirty tool, a so-called *product-process matrix* that can help organisations to revive FM's scope, role and function within their organisation. The tool assists FM organisations to assess its current value position and to outline the desirable value-added position. A comparison of an existing value position and a possible value-added position allows FM practitioners to identify actions for improvements. It should be noted that by FM organisation we mean both internal FM functions and related external FM providers.

The product-process matrix consists of its two building axes, as shown in Figure B26.1. The vertical axis concerns the FM product: What do FM organisations offer to their clients? There are four types of FM products based on the degree of customisation and complexity. The horizontal axis concerns the FM process through which clients interact with FM organisations: How do they collaborate and interact with each other? Here, there are also four categories based on the degree of collaborative relationship between FM organisations and their clients.

The main rationale behind the FM product-process matrix is to underpin an optimal matching of product ('what' is delivered?) and process ('how' is it delivered?). Combining the categories on both axes, there are four value-added positions along the diagonal of FM product and process axes. Each category of FM product requires different types of FM process. This matrix hypothesises that an appropriate matching will deliver added value to the core business and surroundings. The following sections explains the characteristics of each axis.

FM product: what does the FM organisation offer?

FM products cover a wide array of very diverse services, ranging from operational services to managerial expertise, including specialised know-how and knowledge. Some products involve routine, low-skill activities, others are highly

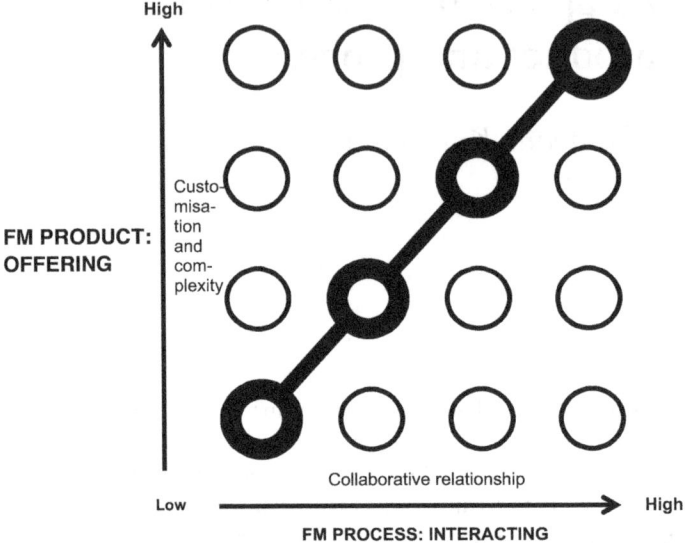

Figure B26.1 The structure of a product-process matrix

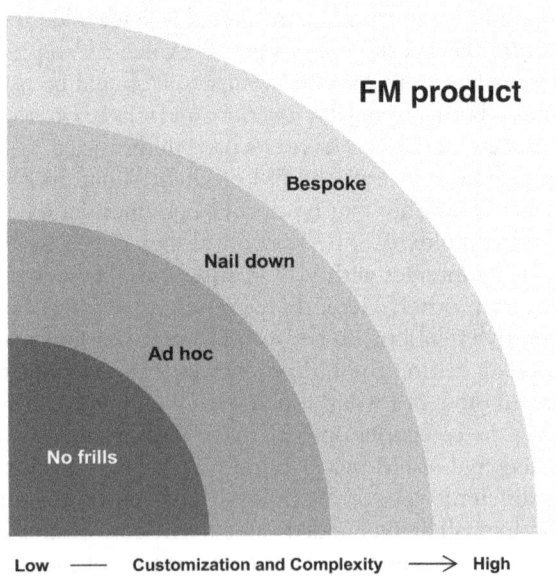

Figure B26.2 Four continuums of FM product

complex, requiring high levels of creativity and strategic thinking. To be able to discuss FM's value position, we make a rough distinction between four FM products based on the degree of the customisation and complexity, as shown in Figure B26.2.

1 **No frills**

This type of FM product concerns commoditised products with a high level of standardisation and few options. It is mostly seen as a stable facilities product for supporting organisational activities. It is mostly routine day-to-day basic operations, creating short-term effects on its host organisation. It is not very strategic, but end users explicitly perceive and comment on the quality and effectiveness of this facilities' product rather than other types of facilities' products. Examples include cleaning and housekeeping, catering, reception and maintenance.

2 **Ad hoc**

This type of FM product includes more complex specifications than the 'no frills product' and offers more options while still being rather operational. It aims to respond to the current organisational strategy and to initiate the organisation capability and capacity. The product focusses on short-medium impacts to its client. Its output is primarily reviewed by organisational business units, who procure according to service level agreements (SLAs) and key performance indicators (KPIs) – for example, the trend of environmental certification in FM practices that attempts to decrease the social pressure on an environmental impact of FM usability and to brand the corporation's reputation in terms of public good.

3 **Nail down**

This product offers a more customised facilities product to ensure the functionality of its client's primary activity. It is very much about facilitating the continuity of an organisation's business processes, reducing the risks of failures or downtime. Because of its more tailor-made product and longer-range effect, it requires FM organisations to be involved in the strategic planning process with its clients. The effects of FM products critically influence the organisational primary activity. An example is cleaning in hospitals or laboratories. Cleaning is still a rather routine-based activity, but for the organisations, hygiene levels are much more critical than in office buildings, where it would probably be a 'no frills' product.

4 **Bespoke**

Bespoke FM products are highly unique products and closely associated with an organisation's business process. Bespoke products contribute the long-term impact and affects organisational bottom lines directly perceived by clients and top management teams. Facilities managers particularly engage in an organisation's strategic planning process. FM organisations and clients share the mutually agreed upon goal – for instance, benefit and risk sharing. Examples include retail outlets that arouse the unarticulated shopper's needs by using disoriented space layout and create unnerved shopping experiences by relaxing shopping environment.

FM process: how clients interact with FM organisations?

Given the four types of FM products, the second axis concerns the question of how these products can be delivered to the client organisation, looking in

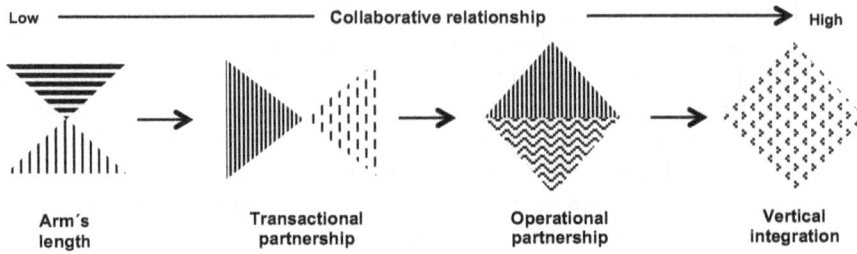

Figure B26.3 An array of FM process

particular at the relationship between the FM organisation and its client. In the process-product matrix, we differentiate between four types of FM delivery processes, based on the level of collaboration, as shown in Figure B26.3. The four types of collaborative relationship are defined by the extent to which they are based on mutual involvements with benefit and trust sharing, mutually agreed upon goals, strategic involvement, input/ output transferring includes data, information, knowledge and innovation.

1   **Arm's length**
    The second category concerns the conventional spot market relationship. There is no risk and benefit sharing equally. Clients focus on cost reduction, and FM providers aim for achieving service level agreements (SLA) and key performance indicators (KPI). Informative data are transferred from the client to FM providers in a top-down manner. FM organisations are not involved in the decision process of FM product specifications with clients. There are no significant differences between the different FM provider's performances.

2   **Transactional partnership**
    With a 'transactional partnership', there is an equal input/output transfer relationship between the FM provider and its client. They tend to mutually share certain economic risks and benefits. However, cost reduction still plays a key role in decision-making process. FM engages in the operational and tactical level in order to align FM working processes with the organisation's primary activities.

3   **Operational partnership**
    In this type of process, an FM organisation plays a key role on a tactical or strategic level. Clients and FM organisation share mutual risk, benefit, trust and commitment. Their relationships move from a spot market-driven relationship towards becoming the preferred partners. Facilities products are embedded in the organisation's primary activities and the client's core business. The degree of knowledge sharing between FM organisations and their clients is high.

4 Vertical integration

In this last category, FM is the primary activity in its client's core business. Clients heavily rely on FM's performance and activities. The scopes of FM activity are not only meeting the client's needs but also meeting the needs of the client's customers. There is the mutually agreed upon relationship between the FM organisation and its host organisation. They share mutual risk, trust, benefit and commitment, including the organisational bottom line. The roles and responsibilities of FM shift away from supportive roles towards becoming an organisational core business. There is only a single FM provider per client because of the close relationship. FM organisation engages in an organisation's decision-making process. At a strategic level, FM and clients jointly create shared knowledge.

Product-process matrix

Combing the discussed categories and axes creates a matrix in which the cells depict the interrelationship between the facilities product and the facilities process of the FM value-added position. Our basic hypothesis is that the cells which match a FM product with a FM process along with the diagonal will have a greater potential to deliver value-added to the stakeholders than the cell that is located away from the diagonal, as shown in Figure B26.4. Furthermore, our

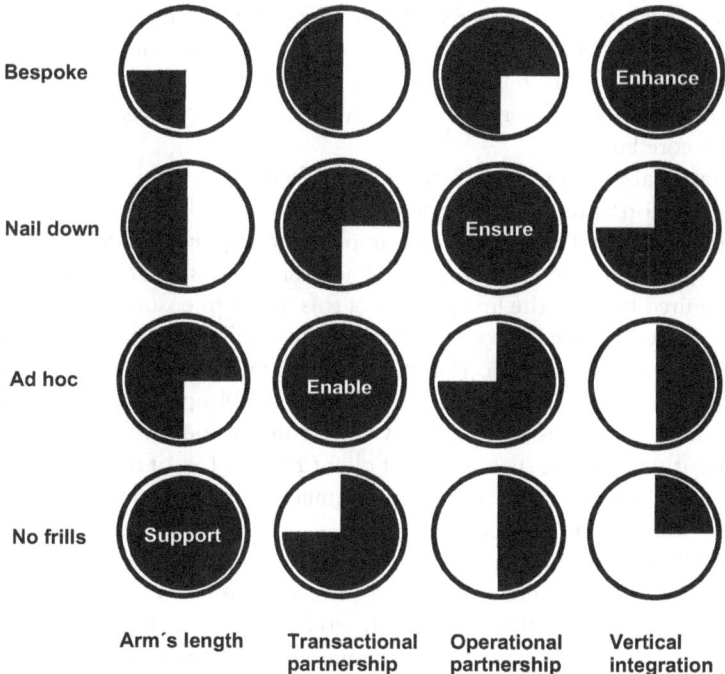

Figure B26.4 Illustrating four value-added positions along the diagonal

hypothesis is that the potential to deliver added value is larger the higher the position is on the diagonal.

However, each position has its own merits and contingencies under specific circumstances for its client's core business. The main rationale behind of the FM product-process matrix is to find an efficient matching of product and process. Matching a given type of FM value-added position with the appropriate FM product and process under the specific conditions is likely to create greater value to the client's core business. FM organisations must align and allocate FM products that meet the needs and requirements of its stakeholder discretion and decision. Based on this matching in the matrix, four value-added positions emerge:

1   **Arm's length + No frills = Support**

In this position, the FM organisation functions as a sort of back office to its host organisation, supporting core business activities, up-and-running on the regular basis without disturbing bottom prices. The downtime of possible hick ups of failures in the FM performance does not critically influence the core business operation. FM does not offer the cutting-edge or high-value products, but it rather cuts cost down and reduces the hassle from primary activity operation.

2   **Transactional partnership + Ad hoc = Enable**

Here the FM performance accommodates the purposeful function of not only supporting core business's activities, but it also enables the current corporate capability and capacity. FM products offer cutting-edge practices and performance with specific timeframe. FM providers need to collaborate with one or more business units to customise the product specification. The occasional downtime would not create major negative impacts on the core business.

3   **Operational partnership + Nail down = Ensure**

This position aims to ensure the constant functionality of an organisation's primary activity. FM may not implement any up-to-date FM practices as quickly as the enabling role. The best in class of FM products is not required because the first priority of this role is to ensure and monitor the working situation of core business activities. FM plays the critical role of hosting the organisation because the FM process involves or is close to core business activities. The downtime of FM operation influences the primary activities tremendously. The core business will experience the negative impacts, such as loss of client royalty, loss of revenue and brand reputation. FM activities and performances are embedded in the organisational business process.

4   **Vertical integration + Bespoke = Enhance**

FM aims to enhance organisation's triple bottom lines by satisfying end users. FM products directly influences the end user's perceptions and experience. Stakeholders and core business require the most reliable and innovative FM products. It influences organisation bottom lines – for example, revenue in the private sector or end user's well-being in the public sector. FM organisation co-creates the FM product's specification with all relevant stakeholders involving client, business units and end users. The downtime of FM products will immediately influence the core business.

Conclusion

To come back to the discussion about whether FM is a strategic discipline or just a cost of doing business, we have argued in this article that the answer is context dependent. From the matrix, it can be argued that a given value-added position has its own merit depending on the host organisational core business' needs and requirements on FM products. FM can certainly be of strategic value, but mostly when it concerns 'bespoke', or – to a lesser extent – 'nail down' FM products. For 'no frills' products, however, it is OK to leave the strategy discussion behind and just have an 'arm's length' relationship, which is mostly about costs. Admittedly, the matrix is a 'quick and dirty' tool – a discussion starter rather than an exact measurement instrument, but its message is important: FM practitioners need to align their offering with the right approach and channel rather than at any price pursuing the attention of the higher managerial level without an obvious cause.

Literature guide

The chapter is based on the following PhD thesis:

Katchamart, Akarapong: Profiling value-added position in FM. PhD thesis 8.2013, DTU Management Engineering. May 2013.

The FM product-process is also presented in the following book chapter and scientific article:

Akarapong Katchamart: Classifying FM value positioning by using a product-process matrix. Chapter 8.2 in Per Anker Jensen and Susanne Balslev Nielsen (eds.): *Facilities Management Research in the Nordic Countries: Past, Present and Future*. Centre for Facilities Management – Realdania Research, DTU Management Engineering, and Polyteknisk Forlag, January 2012.

Katchamart, Akarapong: Mapping value added positions in facilities management by using a product-process matrix. *Journal of Facilities Management*, Vol. 11, No. 3, 2013, pp. 226–252.

Other CFM publications with Akarapong Katchamart as sole author include the following:

Katchamart, Akarapong: Before jumping on the CSR bandwagon: Do we have a parachute? 13th International FM&REM-Congress: "Built Environment", Kufstein, Austria, 27–29 January 2011.

Katchamart, Akarapong: Stakeholder's influence on FM strategy: A case study of an energy complex in Thailand. Chapter 14 in Per Anker Jensen, Theo van der

Voordt and Christian Coenen (eds.): *The Added Value of Facilities Management: Concepts, Findings and Perspectives*. Centre for Facilities Management – Realdania Research, DTU Management Engineering, and Polyteknisk Forlag, May 2012.

Katchamart, Akarapong: Blue collar FM workers as a primary stakeholder: A case study of CSR practices. Proceedings of the 11th EuroFM Research Symposium, 24–25 May in Copenhagen, Denmark. Centre for Facilities Management – Realdania Research, DTU Management Engineering, and Polyteknisk Forlag, May 2012.

Another CFM publication has Akarapong Katchamart as co-author:

Jensen, Per Anker and Akarapong Katchamart: Value adding management: A concept and a case. Chapter 10 in Per Anker Jensen, Theo van der Voordt and Christian Coenen (eds.): *The Added Value of Facilities Management: Concepts, Findings and Perspectives*. Centre for Facilities Management – Realdania Research, DTU Management Engineering, and Polyteknisk Forlag, May 2012.

Use of the model – assessed by Per Anker Jensen

As shown in the table that follows, the FM product-process matrix is particularly useful in two of the five processes presented in part A: strategy development and process optimisation. The model can be used as a mapping and analysis tool, possibly in combination with the models in Chapters B.25, B.27, B.28 and B.29.

Process	Phase							
Strategy development	A	B	C	D	E	F		
Organisational design	A	B	C	D	E	F		
Space planning	A	B	C	D	E	F	G	H
Building project	A	B	C	D	E	F	G	H
Process optimisation	A	B	C	D	E	F		

For strategy development, the tool can be used by managers and staff on the strategic level in FM organisations to map and analyse the company's current need for FM services and arrange the provisions, as well as define and develop strategy plans (phases B–D).

For process optimisation, the tool can be used by managers and staff on the strategic level in FM organisations and their consultants in all phases from evaluation of current performance and improvement potentials to evaluation of new performance and assessment of need for further optimisation, and as a basis for dialogue with internal and external stakeholders (phases A–F).

B.27 Value-Adding Management

Per Anker Jensen

Introduction

In recent years, it has become more and more evident that FM needs to deliver added value to the core business and that it is no longer sufficient only to steer on cost reductions. Many FM organisations have realised that and work on developing new competences and management tools to achieve it.

The purpose of this chapter is to investigate and develop a new management concept for 'Value-Adding Management' (VAM), which can support FM organisations in their attempts to deliver added value in a systematic way. The concept will focus on the effectiveness of FM and is seen as supplementary to internal process management focussing on efficiency of FM. It will address the relationships between a FM organisation and the core business it supports on a strategic, tactical and operational level.

An illustration of how VAM is distinguished in comparison to other forms of management in relation to effectiveness and efficiency is shown in Figure B27.1. If there is a lack of management focus, it is likely that both efficiency and effectiveness is low, which is the situation shown in the bottom left corner called Laissez Faire Management. The situation where the management focus is on optimising efficiency is shown in the bottom right corner and called Industrial Management. This is equivalent to traditional management methods in manufacturing based on Frederick Taylor's so-called scientific management tools like motion and time methods (MTM), and modern concepts like lean or agile management are typical examples on management with a dominating focus on efficiency.

The opposite situation where the management focus is on optimising effectiveness is shown in the top-left corner and called Preparedness Management. A fire brigade is an extreme example of this situation, where one has an organisation standing by in case of the occurrence of a certain undesirable event, but any management concept where high effectiveness has priority whatever the cost is in this category. VAM is placed in the top-right corner where both effectiveness and efficiency have high priority.

I have earlier investigated the organisational relationship between FM and core business on the strategic and operational levels with inspiration from theory on governance and forms of coordination. This is presented further in Chapter B.32.

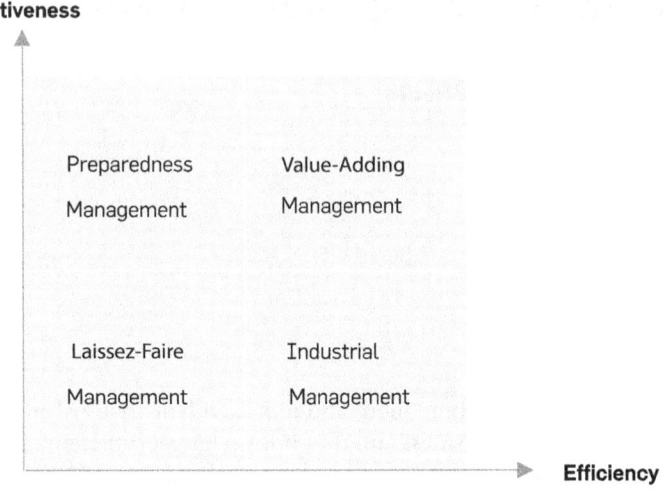

Figure B27.1 Value Adding Management compared with other forms of management

The conclusions are that for decision-making related to strategic FM concerning common corporate capacity and infrastructure, it is important to create a close collaboration and alignment between the FM organisation and the core business to achieve the necessary business orientation. Such a collaboration could take the form of a coalition managed by a forum of representatives from FM and the different parts of the company. In the case of conflicts and disagreements, the company board of directors could act as a steering committee.

As a contrast for FM provisions with a differentiation in relation to various internal users, de-centralised decision-making seems to be the obvious solution. That is particularly the case where the quality of the provision is easily defined and understood by both parties, and in those cases, price seems to be the best form of coordination and a service orientation is essential. Examples of this could be cleaning, catering, internal removals, hiring of conference rooms and procuring of standard products.

For more complex provisions with the need for dialogue about specific customisation, more centralised decision-making may be needed involving negotiation between managers at some level. This can, for instance, be department managers or managers of business units who act as customer representatives for their units and their specific needs. Space management issues, like rebuilding projects and workplace design, could be typical examples. Here there is a need for the FM organisation to have a customer orientation.

The distinction between the management levels strategic, tactical and operational is in accordance with the European FM standards (see Chapter B.15). Based on the aforementioned, it seems essential that the relationship management in VAM is differentiated on the three levels as shown in Table B27.1. Business

Table B27.1 Differentiation of relations in Value-Adding Management

Level	Demand side	Relationship focus	Coordination form
Strategic	Client	Business orientation	Coalition
Tactical	Customer	Customer orientation	Negotiation
Operational	End user	Service orientation	Price per order/service charge

orientation means that considerations for the whole corporation is in focus, and this calls for joint decision-making involving all main stakeholders at management level, which can take the form of a coalition. Customer orientation means that the specific needs of each business unit are in focus and this calls for a bilateral negotiation and decision-making. Service orientation means that individual users' needs are in focus, and the services are either provided based on price per order – for instance, catering and transportation – or based on a service charge, which for instance, can be part of internal rent or similar, such as cleaning and security.

One of the most challenging aspects of VAM is how to measure the added value. This will probably have to include both qualitative and quantitative measurements. A possible method could be the Balanced Scorecard developed by Kaplan and Norton, which is probably the most commonly used management measuring method besides financial measurements. How to measure added value is treated further in Chapter B.29. In the following, I will present a case from Lego, which shows how they have worked with FM in a way that to a high degree resembles VAM, and they have even developed a method to measure added value or value add as they named it. It must be stressed that the information was collected around 2010 and does not necessarily represent the current situation in the company.

Case from Lego

Lego is a Danish family owned company producing construction toy products for the global market. Lego's headquarters is placed in Billund, but they have production facilities and sales offices around the world. The Lego group has approximately 9,000 employees. FM in Lego is a part of Lego Service Centre (LSC), which is an integrated business unit encompassing support services such as information technology (IT), human resources (HR), indirect procurement and reception besides FM. The FM unit is responsible for all Lego's facilities around the world.

Lego uses the Balanced Scorecard as a corporate management tool. For FM, they have developed a strategic map, as shown in Figure B27.2, where

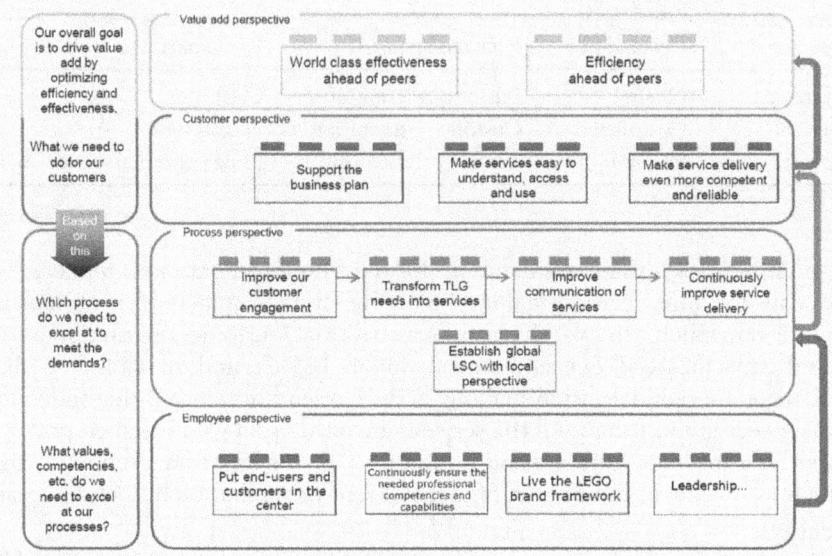

Figure B27.2 Lego's facilities strategy map 2010

the financial perspective at the top level from the original strategic mapping methodology by Kaplan and Norton is replaced by a value add perspective divided into effectiveness and efficiency.

Value add can, according to Lego, be divided into financial and non-financial value – financial value examined by cost reduction and non-financial value by volume, quality and flexibility as specified next. It can also be shown as CO_2 emission reduction, environmental portfolio and green account. An added value report is a supplementary part of the financial report, which is delivered to LSC's client and customers. The objective of FM is to deliver minimum 5% value adding every year. This is measured by the so-called value add equation:

$$\text{Value add} = \text{Volume} * \text{Quality} * \text{Flexibility/Cost.}$$

Volume represents the level of scalability and is calculated as the number of standard services (part of the service catalogue) delivered. *Quality* is measured in two ways. The user perceived quality is measured by the use of surveys, which are sent to randomly selected users. But the quality can also be documented with quantitative metrics where the incident rate as a percentage of all services delivered can be calculated. *Flexibility* is measured in surveys where the buyers' perception of LSC's flexibility is evaluated. Only initiatives which are initiated by LSC and recognised as adding value by the customers benefitting from the initiatives are accepted as value add and can be included in the calculation. The value equation is seen as a performance measurement tool and a basis for dialogue with internal stakeholders, but also as a tool internally for the staff in the FM unit to put focus on why they are there.

The relationship between LSC and its internal stakeholders is shown in Table B27.2. It distinguishes between the same three levels of management and the same types of relationships that are included in VAM (see Table B27.1).

LSC posits itself as a middle management moderator between a strategic level and an operational level. To maximise delivered service values, LSC needs to juggle the varying interest of the different stakeholders. FM as a part of LSC has to compromise needs of end users, customers and top management, thus it is necessary to understand the needs by creating communication channels to align its service delivery with their expectations.

Table B27.2 Relations between LSC and Lego's core business

Stakeholder	Communication channel	Target group	Focus area
Client	Leadership team survey and meeting	Top-40 management level included the vice president	Where are we? Do they see LSC as added value to Lego business? What are we doing with added value? Is LSC a good partnership with Lego?
	LSC Facility Committee meetings	Comprising of CFO, head of Lego Corporate Centre, head of Global Supply Chain, head of LSC and by invitation head of Marketing and Products and head of Customer and Education Division	Prioritising LSC services and makes decision across the board Aligning LSC services with business process
Customer	Customer meetings	Customers are director level and above	Agreement on key performance indicator (KPI) and service level agreement (SLA)
	Customer survey		Do they understand/ know LSC services?
End users	User survey	Users are everyone below director level	Do they understand/ know LSC services?

The management of LSC participates in an annual meeting with Lego's top management – the leadership team – to evaluate performance and discuss development plans. In order to align strategic management decisions between top management and FM on a continuous basis, Lego has established the LSC Facility Committee with the main focus on the three aspects: projects, capacity and competency. The meetings are held every six weeks. In addition, the service levels are negotiated and decided bilaterally with the management of each business unit as customers. LSC also measures their performance based on satisfaction surveys with regular intervals. These surveys are differentiated in relation to the client, customers and end users as shown in Table B27.2.

Conclusion

This chapter has presented a concept for VAM. The case study of Lego represents an exemplary case of an organisation actually practising VAM without using this exact term. The management in LSC is actively working with adding value by FM to the core business. They have established procedures and communication channels for defining what can be accepted by the core business as representing value-adding by FM. The communication is differentiated between the client at the strategic level, the customer at tactical level and the end users at operational level in line with the European FM standards and the proposed VAM concept.

The Facility Committee in Lego is a clear example of a coalition between FM and top management aiming at joint decision-making about strategic investments with an overall business orientation. The specific FM service levels are negotiated and agreed bilaterally with each business unit and thus differentiated with a customer orientation. The individual services are delivered to the end users with a service orientation. On each level, the performance is measured by differentiated satisfaction surveys by regular intervals. Lego has even developed a method to quantitatively measure the value adding.

Literature guide

The chapter is based on the following conference paper and book chapter:

Jensen, Per Anker and Akarapong Katchamart: Value added management: A new facilities management concept. *EFMC2011 Conference and Research Symposium, Vienna, 2011.*

Jensen, Per Anker and Akarapong Katchamart: Value adding management: A concept and a case. Chapter 10 in Per Anker Jensen, Theo van der Voordt and Christian Coenen (eds.): *The Added Value of Facilities Management: Concepts, Findings and Perspectives.* Centre for Facilities Management – Realdania Research, DTU Management Engineering, and Polyteknisk Forlag, May 2012.

Use of the method

As shown in the table that follows, Value-Adding Management is particularly useful in two of the processes presented in part A: organisational design and process optimisation. The method can be used as an analysis tool, possibly in combination with the models and methods in Chapters B.25, B.26, B.28 and B.29. The case from Lego can be used as inspiration.

Process	Phase							
Strategy development	A	B	C	D	E	F		
Organisational design	A	B	C	D	E	F		
Space planning	A	B	C	D	E	F	G	H
Building project	A	B	C	D	E	F	G	H
Process optimisation	A	B	C	D	E	F		

For organisational design, the method can be used by managers and staff on the strategic level in FM functions and their consultants to evaluate the current organisation, to prepare proposals for a changed organisation and evaluating the need for new knowledge and competences, and as a basis for dialogue with top management and other internal stakeholders (phases B–D).

For process optimisation, the method can be used by managers and staff on the strategic level in FM functions and their consultants in all phases from evaluating current performance and improvement potential to evaluate new performance and assess the need for further optimisation, and as basis for dialogue with internal and external stakeholders (phases A–F).

B.28 How IT provides added value to FM

Poul Ebbesen

Introduction

Assessing the added value from the effort of implementing information systems (IS) supporting FM processes is associated with major challenges. It is often unclear what added value is expected and what part of the supply chain of FM deliveries that benefits from the IS. One reason for this might be that the concept of value is not well defined and as a consequence can be difficult to use as the sole parameter. Furthermore, it is often not understood very well how the parts in the supply chain of FM deliveries are interconnected. Therefore, a general method for assessing the added value of IS in FM is proposed.

The method is based on the concept of added value as described in Chapter B.27. This implies a fundamental distinction between efficiency and effectiveness. Another fundamental distinction is made between use value and exchange value (economic value). Relations between these concepts are illustrated in Figure B28.1.

In relation to IS, interoperability can contribute significantly to efficiency. Interoperability is the ability to exchange data between applications in a seamless manner in a coherent IS. The concept of functional affordances is another aspect of IS. Functional affordances are the potential uses of an IS based on the user's interpretation of the use qualities that the IS offers.

To achieve added value from an IS, it is most often necessary to invest in purchasing and installation of hardware and software, development of customised solutions, collection and storage of data and training of staff. Figure B28.2 illustrates an ideal development of costs and use value over time.

The baseline for use value can be specified in a service level agreement (SLA). The use value of the service can for instance be measured by key performance indicators (KPI) with a minimum level of customer satisfaction. An increase in use value will occur if the customer satisfaction over time gets higher than the specified minimum level of customer satisfaction. This means that added-use value is created. A cost reduction of the service occurs if the cost of the service goes down below the baseline without lowering the customer satisfaction below the specified minimum level. The curve at the bottom of Figure B28.2 shaped as a hump represents the investment in the IS implementation. It can be seen as the sacrifice or effort of IS implementation.

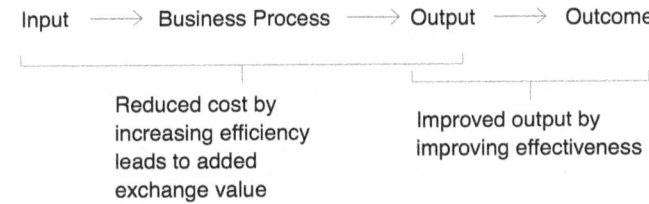

Figure B28.1 Added exchange value and added-use value

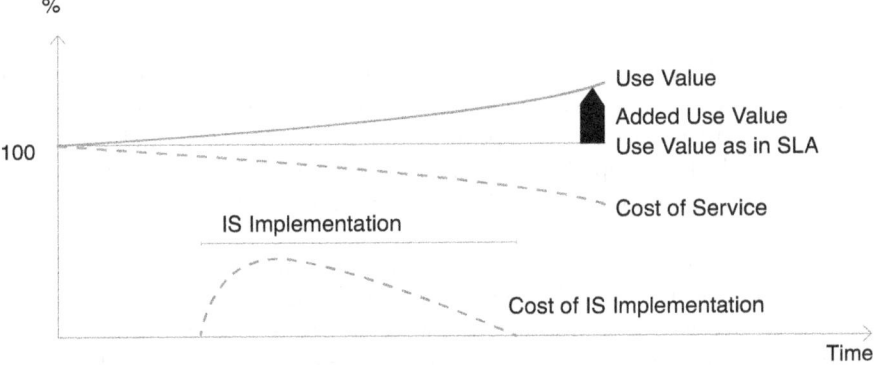

Figure B28.2 Ideal relative development over time in use value, cost of service and cost of IS implementation

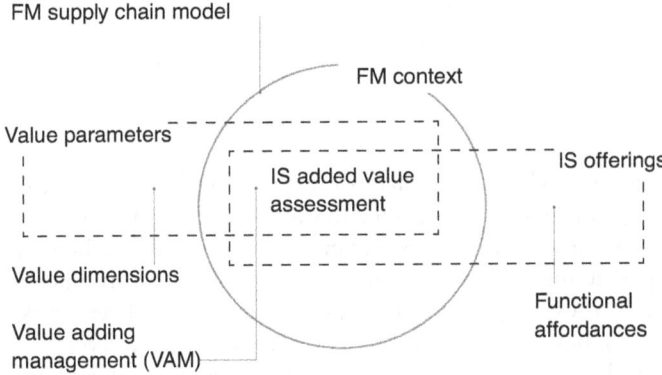

Figure B28.3 Elements constituting the IS added value assessment method

The proposed method combines existing concepts and models from different disciplines, which is illustrated in Figure B28.3. Firstly, value parameters (value dimensions) suitable for describing the value of business processes were found in the existing literature on value, IS and FM. Secondly, an overall concept

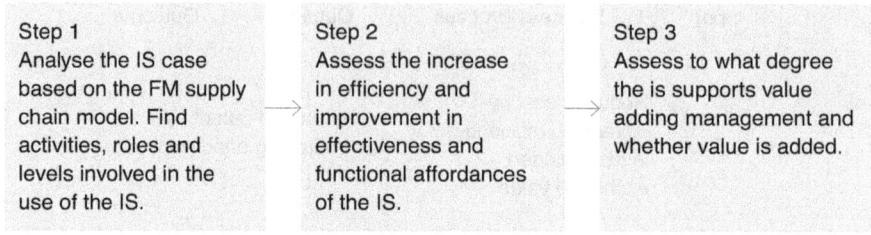

Step 1	Step 2	Step 3
Analyse the IS case based on the FM supply chain model. Find activities, roles and levels involved in the use of the IS.	Assess the increase in efficiency and improvement in effectiveness and functional affordances of the IS.	Assess to what degree the is supports value adding management and whether value is added.

Figure B28.4 Three steps in assessing the value of IS supporting FM

(functional affordances) for describing, what IS can offer in relation to the value dimensions was chosen. Next, a model describing the FM context (FM supply chain model) based on the European FM standard was found, and, finally, the model described in Chapter B.27 for assessing the value of management based on the value dimensions was included (VAM). See Figure B28.6 under the case that follows.

The method consists of three steps as illustrated in Figure B28.4. In the first step, the IS case is analysed using the FM supply chain model. Activities, roles and levels involved in and interacting with the IS are identified. In the second step, increase in efficiency and improvement in effectiveness as a result of the IS implementation is assessed. This is done using the definitions of efficiency and effectiveness illustrated in Figure B28.1. Furthermore, the functional affordances of the IS, which contributes to increase in efficiency and improvement in effectiveness are identified. Finally, in step 3, the degree or level of VAM is assessed.

Case

Use of the method is exemplified using a case study. The organisation is a large European airport. The specific IS was put into use in 2013 and supports cleaning in the airport. The IS delivers information about the frequency of use of specific intensely used rooms. Sensors detect when a person enters into one of these rooms, and the system can thereby keep track of, how many people have used each room. Furthermore, a use response system has been installed in each of the room, enabling users on their way out of a room to report back about their experience of using the room; whether it was good, OK or bad. If responding 'bad', the user is asked to report back which specific issues caused the bad experience?

Data from the sensors and from the use response system is presented on a monitor in a simple way to the dispatcher of the external cleaning company. The dispatcher's role is to coordinate the cleaning process based on the information on the monitor. When a specific number of persons have passed the sensor in a room, a field on the dispatcher's monitor turns from green to red, and if a specific percentage of users of a room find the experience of using the room bad, another field turns from green to red. In each case, the dispatcher then can send cleaning personnel to this specific room to see, whether cleaning is necessary. If a user reports back about a specific issue that needs to be improved, the cleaning manager can also send personnel to the room to deal with the issue.

Before the IS was taken into use, cleaning in these room were done on a regular basis, e.g. every two hours. Whether there actually was a need for cleaning, or whether specific issues needed to be handled was not part of the cleaning procedures. As a consequence of implementing the IS, cleaning in these rooms is now demand driven; based on use frequency and user experience responses. The external cleaning company are, according to their contract, required to use the IS. There are no explicit SLA or KPI sections in the contract between the cleaning company and the airport, but the contract requires the cleaning company to deliver cleaning according to normal standards. The contract does, however, require the cleaning manager from the cleaning company to send out cleaning personnel, when and where it is needed, based on information from the IS, and to respond to specific issues reported by users.

As part of an international benchmarking of airports, users of the airport have the last seven years been asked quarterly about their experience of using the facilities, including their experience with the level of cleaning in these rooms. In general, the satisfaction level has increased every year. The introduction of this IS and the change in procedures may have contributed to the increase of the satisfaction level, but also refurbishments and other improvements of the rooms during the same period may have had an impact. Nevertheless, the introduction of this IS and the changes in the work procedures is aligned with the goals and strategies defined by top management, which includes aiming at being seen by users as the best airport in Europe.

Based on this case, it is now shown how the three steps in the method can be used to assess the value added by implementing IS supporting FM into an organisation.

Analysing the IS case using the FM supply chain model (step 1)

The analysis refers to the numbers in brackets in Figure B28.5. The IS reports frequency of use detected by sensors, and level of experience and issues reported by users (1). On the supply side, the information from the IS is used by the dispatcher (2) from the external cleaning company

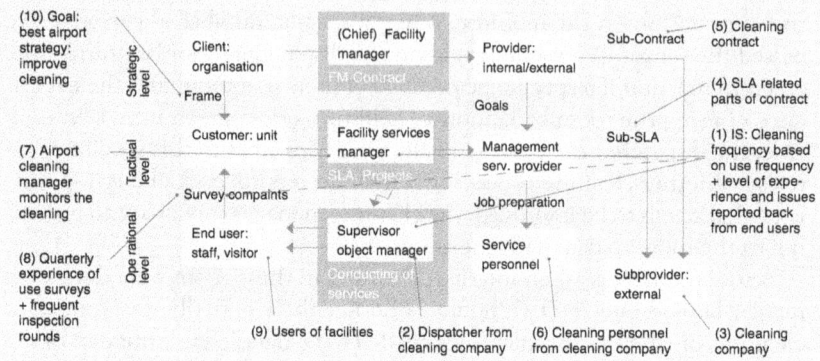

Figure B28.5 Analysis of IT case by use of the FM supply chain model

(3) to manage the cleaning, including the cleaning frequency. The SLA (4) related part of the contract (5) between the cleaning company and the airport requires the cleaning manager from the cleaning company to send out cleaning personnel (6) when and where it is needed, based on information from the IS, and to respond to specific issues reported by users. On the demand side the manager responsible for cleaning at the airport (7) monitors the cleaning based on information from the IS, supplemented with occasional inspection rounds in the facilities (8). By use of surveys (8), the user's (9) experience of the facilities is investigated quarterly. The surveys, in regard to cleaning, show an increase in the satisfaction level among the users. The use of the IS to support the cleaning business process is aligned with the strategy of improving cleaning and thereby contributes to reach the goal, set by top management, of being the best airport (10).

Assessing efficiency, effectiveness and functional affordances of the IS (step 2)

The quality of the cleaning (output in Figure B28.1) has, according to the cleaning manager, improved since the introduction of the IS. Therefore, the effectiveness of the cleaning has improved. At the same time, the experienced level of cleaning (outcome in Figure B28.1) has increased according to the quarterly survey results. The improvement in effectiveness has led to improvements in the output (and outcome), thereby leading to added-use value (see Figure B28.1).

Since the IS was introduced the frequency of cleaning has dropped, while the level of cleaning has not dropped. The efficiency of the cleaning therefore has been improved. The airport pays the cleaning company the same for the cleaning, compared to before the IS was introduced. The cleaning company may have reduced their cost as a consequence of increase in efficiency and thereby achieved an added exchange value, but this is not the case for the airport (at least not at the time of this case study, but this may change over time, e.g. at renewal of the contract.)

The functional affordance of the IS, which has increased efficiency, is the delivery of real time user frequency information on the monitor. This functional affordance has made it possible for the dispatcher to send out cleaning personnel, when a certain number of persons have used a room. The functional affordance of the IS, which has improved effectiveness, is the delivery of user response information on experience of use and on specific issues. This functional affordance has made it possible to send out cleaning personnel, when specific issues must be dealt with. The fact that the cleaning manager hired by the airport uses the IS to monitor the cleaning process can be seen as a functional affordance, which has improved interoperability. The IS facilitates sharing of data about the cleaning process.

Assessing how the IS supports VAM and whether value is added (step 3)

The IS supports effectiveness. It delivers information, so the cleaning manager can better initiate the right cleaning, thereby improving the effect of the cleaning (added-use value). The IS also supports efficiency. The cleaning manager can better initiate cleaning, when it is needed, and thereby reduce the resources spent on cleaning (added exchange value). Supporting both efficiency and effectiveness in the management of the process, the IS therefore supports VAM (see Figure B28.6).

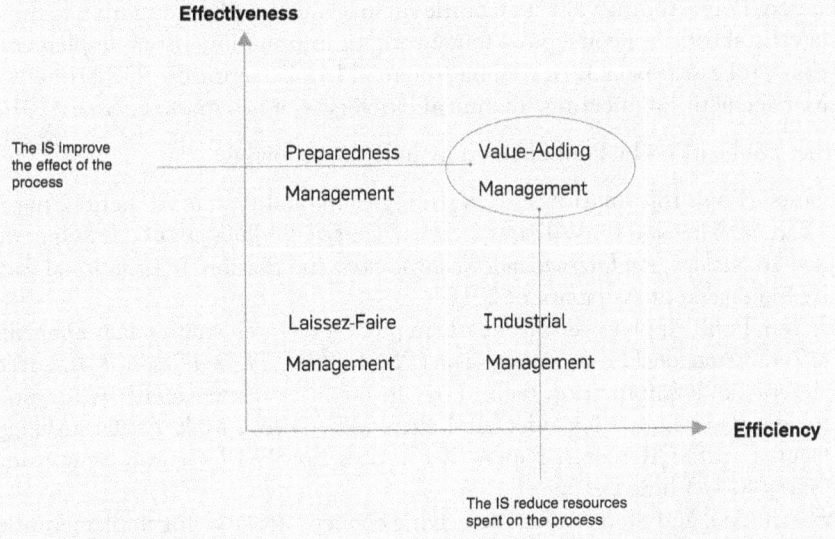

Figure B28.6 The IS supports both efficiency and effectiveness in the management process

Conclusion

This analysis cannot stand alone. It does not include an assessment of the quality of the management, the work process or the IS. This analysis only helps clarify, whether the basic managerial and technological elements needed to achieve VAM are present. The case also illustrates that a well organised management setup is required to gain value from IS. Furthermore, it is illustrated that implementing IS includes both organisational and technological changes.

The use of the IS as described definitely adds value to cleaning, being a secondary and supporting process. Because of the increase of the user experience of the cleaning level, which is aligned with the strategy of improving cleaning in order to become the best airport, the use of the IS also adds value to a primary process of the organisation.

Literature guide

This chapter is mainly based on the following conference paper:

Ebbesen, Poul and Per Anker Jensen: Assessing the added value of information systems supporting facilities management business processes. Paper in Susanne Balslev Nielsen, Per Anker Jensen and Rikke Brinkø (eds.): *Research Papers for EuroFM's 16th Research Symposium at EFMC2017, 25–28 April in Madrid, Spain*. EuroFM, Centre for Facilities Management – Realdania Research, DTU Management Engineering, and Polyteknisk Forlag, April 2017.

The development of the method was part of Poul Ebbesen's PhD study and the full results of this can be found in his PhD thesis:

Ebbesen, Poul: Adding value to facilities management with information technology. Establishing a conceptual framework for information system implementation processes in facilities management. PhD Dissertation. Department of Management Engineering, Technical University of Denmark. February 2016.

Other publications by Poul Ebbesen include the following:

Ebbesen, Poul: Information system strategies in facility management. Chapter 17 in M. May and G. Williams (eds.): *The Facility Manager's Guide to Information Technology, An International Collaboration*, 2nd Edition. International Facility Management Association, 2017.
Ebbesen, Poul: Applying displays early in process research studies. *Paper presented at 7th International Process Symposium (PROS) 2015, Helona Resort, Kos, Greece.*
Ebbesen, Poul: Information technology in facilities management: A literature review. Paper in K. Alexander and I. Price (eds.): *People Make Facilities Management*. EuroFM Research Papers 2015. 12th EuroFM Research Symposium, Glasgow, 1–3 June 2015.
Ebbesen, Poul and Sten Bonke: Identifying concepts for studying implementation of information technology in facilities management. *Proceedings of CIB FM Conference, Technical University of Denmark, 21–23 May 2014.*

Ebbesen, Poul, Jan Karlshøj, Sten Bonke and Per Anker Jensen: Information system strategies in facilities management: Based on five process studies. Paper in Per Anker Jensen (ed.): *Facilities Management Research and Practice – Does FM Contribute to Happiness in the Nordic Countries.* Proceedings of CFM's Second Nordic Conference, 29–30 August 2016. Centre for Facilities Management – Realdania Research, DTU Management Engineering, and Poly-teknisk Forlag, August 2016.

Ebbesen, Poul and Giulia Nardelli: Formal control and scope in information system projects in facilities management: A process perspective. *The 38th Information Systems Research Conference in Scandinavia, IRIS 38, Oulu, Finland, 9–12 August 2015.*

Ebbesen, Poul: Strategisk styret digitalisering i FM-branchen and Digitalisering af bygningsdrift. Chapters 1 and 4 in *Facility Management som digital forandringsagent,* 1st Edition, Dansk Facilities Management netværk, 2018.

Use of the method – assessed by Per Anker Jensen

As shown in the table that follows, the method is especially suitable for use in one of the five processes presented in part A: process optimisation. The method is particularly suitable for optimising processes involving implementation of IS. The model can be used as an analysis tool and for decision support in such processes. The method is closely related to the more general methods for analysis of creation of added value described in the Chapters B.27 and B.29.

Process	Phase							
Strategy development	A	B	C	D	E	F		
Organisational design	A	B	C	D	E	F		
Space management	A	B	C	D	E	F	G	H
Construction projects	A	B	C	D	E	F	G	H
Process optimisation	A	B	C	D	E	F		

In process optimisation, the method can be used by managers and staff at a strategic level in FM functions and their consultants and service providers from the initial phase of evaluation of current performance and improvement potentials, and the subsequent phases for assessing the needs for further optimisation (phases A–F).

B.29 FM as creator of added value

Per Anker Jensen

Introduction

In recent years there has both in practice and research been an increased focus on how FM can create added value for organisations. Within research this has resulted in development of a number of conceptual models and tools as well as collection of much empirical information. However, the practical application of this knowledge has been limited and turned out to be difficult. The reasons seem to be that the different models are to complex and that a common terminology and a clear operationalisation of input-output-outcome relationships is lacking. Therefore, we have as part of the work on a new book developed a simpler model for creation of added value of FM and the related Corporate Real Estate Management (CREM) with the aim to provide a better basis for application in practice. At the same time, we have aimed at creating an integration of knowledge from FM and CREM so that we establish a common foundation that utilises the strong aspects of these closely related disciplines.

One of the existing conceptual models is the FM Value Map, which was developed by me approximately ten years ago and is presented in Chapter B.25. Around the same time there were within research on CREM developed similar conceptual models both at Delft University of Technology in the Netherlands and at Helsinki University of Technology (today part of Aalto University) in Finland. Since then., some of these models have been further developed and supplemented by new ones. The models were an important basis for a former book on FM and added value from 2012, where they were compared and analysed for strong and weak aspects. One of the conclusions was that they were static and that there is a need for more dynamic models to contribute better to management of the process of creating added value.

Both the FM Value Map and other models include a simple process model based on input → throughput → output – even though used in different ways. A closer analysis also revealed that they all implicitly include a cause-effect relationship with great similarities based on interventions as the cause contributing to added value as the effect. This led to the new model for Value-Adding Management (VAM) having the following general process model as a starting point:

Input → Throughput → Output → Outcome → Impact = Added Value

By combining the general process model with the cause-effect model and including Value-Adding Management as the intermediary between cause and effect we can define the generalised Value-Adding Management process model:

Intervention → Management → Added Value

This is in line with the understanding we arrived to in the book from 2012 that management is a prerequisite for the implementation of interventions in FM/CREM can lead to the creation of added value for the organisation. The model can also be expressed as follows:

Decision on type of change → Implementation → Outcome/Impact
And also: What → How → Why.

What is the kind of change and the improvement FM/CREM intends to make to add value; *how* is the way FM/CREM manages the change and implements the improvement, and *why* is the benefit that the core business organisation is expected to achieve – i.e. the positive outcome of benefits versus sacrifices in terms of costs, time and risks.

In the following, the three elements, Interventions, Value-Adding Management and Added Value Parameters, which form the VAM model, are presented briefly.

FM/CREM interventions

We divide FM/CREM interventions in six types.

1   *Changing the physical environment:* This typical include relocation, new building, rebuilding, refurbishment, change in workplace layout and introduction of new design, for instance as part of corporate branding.
2   *Changing facilities services:* This concerns the operational FM activities and includes developing service offering to the user, for instance, new food concept in the canteen, changes in cleaning level or introducing a new user interface like introducing an IT-based help desk.
3   *Changing the interface with core business:* When organisations reach a certain size and complexity, FM/CREM are typically established as separate functions or departments. The interface between the core business and FM/CREM is defined specifically in each organisation and is not static. If the FM/CREM function is successful, it will in many cases get the opportunity to increase its area of responsibility. This is often part of a centralisation of the responsibility from several parts of the core business organisation to the FM/CREM function, thereby creating opportunities for economies of scale.
4   *Changing the supply chain:* FM is in most cases organised as a mixture of an in-house FM function and a number of external providers of facilities services, which constitutes a FM supply chain. Changes in the supply chain with

outsourcing or insourcing primarily include changes in the delivery process, but they often also have consequences for the incentives for the different parties and the management of the mutual relationships between the parties.

5   *Changing the internal processes:* This typically concerns increasing the efficiency of operational processes within a specific organisation without necessarily changing, neither the product, nor the supply chain. The organisation can be in-house or an external provider. Within management theory and practice there are a number of concepts aimed at increasing productivity and process efficiency, for instance Total Quality Management, Business Process Re-engineering, Benchmarking and Lean Management. Typical elements in such concepts are eliminating waste, implementing new technological solutions and optimising the work flow.

6   *Strategic advice and planning:* These are essential elements in the strategic and tactical activities of FM and CREM. The areas for strategic advice and planning can cover many different aspects, and they will typically change over time according to what is of strategic importance for the company. A typical area of strategic advice to top management concerns the development of a long-term strategy for the corporate property portfolio. Another typical area is investment planning and feasibility studies for building projects.

Value-Adding Management

The term 'Value-Adding Management' is widely used in business and management literature. The industrial consultant Carlo Scodanibbio even calls VAM the guiding light for the year 2000 industries. In relation to FM and CREM essential aspects of VAM are strategic alignment between FM/CREM and core business, stakeholder management and relationship management as part of the implementation of changes. Here we will solely focus on strategic alignment.

Aligning implies moving in the same direction, supporting a common purpose, being synchronised in timing and direction and being appropriate for the purpose. FM/CREM can only create added value, when they support the corporate objectives. FM/CREM interventions should not only be checked on its impact on FM/CREM performance and organisational performance but also on its impact on attaining organisational goals. A better performance does not per definition deliver added value. For instance, if an FM intervention results in a higher ranking on 'green buildings' but the organisation was fully satisfied with the original ranking, this higher ranking does not add any value to the organisation.

Added value parameters

Based on the existing conceptual models, we have compared the different value parameters included, and we decided to select 12 parameters. The whole

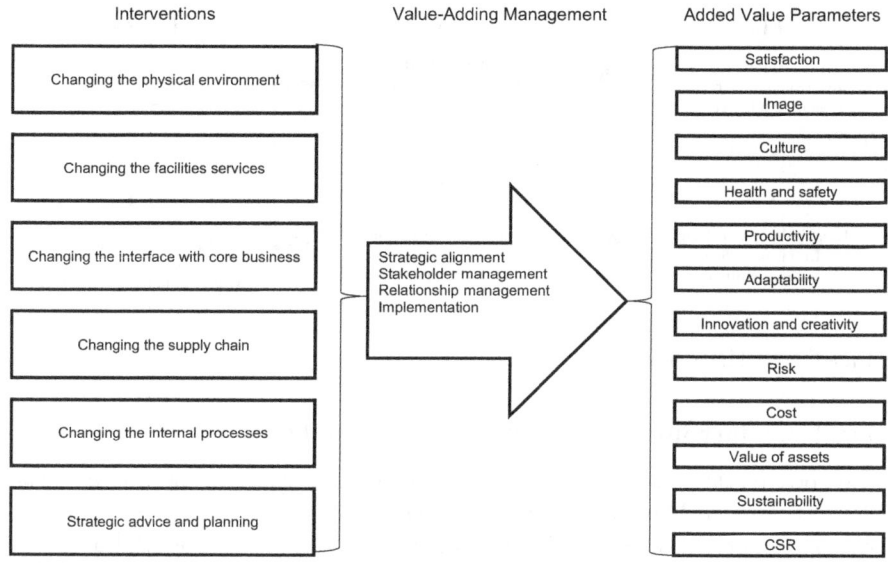

Figure B29.1 VAM model with 6 types of interventions and 12 added value parameters

Value-Adding Management model with the six types of interventions and the 12 added value parameters is shown in Figure B29.1.

The extended VAM model

To make the VAM model for FM/CREM more instrumental and useful as decision support and management tool, we have extended the initial very simple model to include the often used quality circle Plan, Do, Check and Act, see Figure B29.2. The cyclic character underlines that VAM is or should be a continuous process.

A central element in the model is that the evaluation in the Check-phase both include an evaluation of FM performance (output) and an evaluation of the impact of the changed FM performance has on the company's organisational performance (outcome) as a basis to assess, whether added value is actually created. In this context it is important that it as part of the overall evaluation is checked, whether the organisational objectives are met, whether the interventions result in synergy, for instance by supporting more than one value parameter, whether there is a conflict between different results, and whether the results seen as a whole are reasonable considering the cost for relevant stakeholders. The evaluation of the realised output/outcome/added value can be a starting point for starting new interventions.

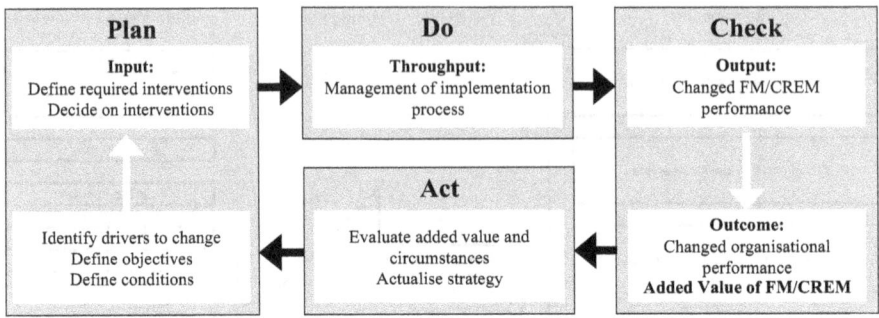

Figure B29.2 The extended VAM model

Supplementary information on the mentioned books

The book *Facilities Management and Corporate Real Estate Management as Value Drivers – How to Manage and Measure Value Adding* was edited by me together with Theo van der Voordt and published in 2017. This book was a follow-up to the book *The Added Value of Facilities Management – Concepts, Findings and Perspectives* from 2012. Both books are based on the work in a research group under EuroFM, which I have chaired since 2009. Furthermore, both books are anthologies with contributions from a large number of authors, who mostly are researcher from different European countries. The newest book also includes 12 interviews with FM/CREM-practitioners on how they work to create added value.

Literature guide

The chapter is presented in full details in the following book:

Jensen, Per Anker and Theo van der Voordt (eds.): *Facilities Management and Corporate Real Estate Management as Value Drivers: How to Manage and Measure Adding Value*. Oxfordshire: Routledge, 2017.

In addition, the model is presented in the following conference paper and scientific article:

Jensen, Per Anker and Theo van der Voordt: Towards an integrated value adding management model for FM and CREM. *Proceedings from CIB World Congress 2016, Tampere, 31 May–3 June 2016.*
van der Voordt, Theo, Per Anker Jensen, Jan Gerard Hoendervanger and Feike Bergsma: Value adding management of buildings and facility services in four steps. *Corporate Real Estate Journal*, Vol. 6, No. 1, 2016, pp. 42–56.

The earlier book is:

Jensen, Per Anker, Theo van der Voordt and Christian Coenen (eds.): *The Added Value of Facilities Management: Concepts, Findings and Perspectives.* Centre for Facilities

Management – Realdania Research, DTU Management Engineering, and Polyteknisk Forlag, May 2012.

Use of the model

As shown in the table that follows, the model is particularly useful in two of the five processes presented in part A: Strategy development and Process optimisation. The model can be used at management and analysis tool, possible in combination the models and methods presented in the former four chapters.

Process	Phase							
Strategy development	A	B	C	D	E	F		
Organisational design	A	B	C	D	E	F		
Space planning	A	B	C	D	E	F	G	H
Building project	A	B	C	D	E	F	G	H
Process optimisation	A	B	C	D	E	F		

For strategy development, the model can be used by managers and staff on the strategic level in FM functions and their consultants to evaluate the current strategy, define strategic goals and develop, implement, follow-up and re-assess the strategic plans (phases B–F).

For process optimisation, the model can be used by managers and staff on all levels in FM functions and their consultants from evaluation of current performance and improvement potential to evaluation of new performance and assessing the need for further optimisation, and as a basis for dialogue with internal and external stakeholders (phases A–F).

B.VI

FM organisation and development

B.30 Development of the integrated FM function

Per Anker Jensen

Introduction

In a research project about the Danish Broadcasting Corporation, DR, as building client and building operator over time, I investigated how these functions, which today are seen as part of FM, have emerged historically. I followed the development from the very start, when DR was established as the Danish State Radio in 1925, through rapid expansion after the introduction of TV in the 1950s, stagnation around the end of monopoly on national TV in 1988 and until the recent establishing of the multimedia centre DR City (DR Byen) as new headquarters. DR is a state-owned company, which is obliged to submit their archives to the Danish national archive institution, and the research included archive studies both at the national archives and at DR's own library, including scrutinising all DR's annual reports and anniversary publications. In addition, I conducted interviews with a number of key persons.

FM related functions in DR are only to a limited degree portrayed in the written sources particularly for the earlier decades. An exception is a chapter called '24 Hours in the Bird's Nest' from DR's 15-year anniversary in 1940. This was at the end of the period, where DR was accommodated in its first self-owned building nicknamed the Bird's Nest. This was a building connecting the old Royal Theatre building with a neighbouring building crossing a road, and it included an extra theatre hall for the Royal Theatre besides DR's facilities. Soon after the anniversary, DR moved to the new Radio House. In that chapter, the serving spirits came out of the darkness. The following is a short excerpt:

24 Hours in the Bird's Nest

> *Just a single half hour during the night time does the Bird's Nest sleep . . . Only the steps of a lonely man can be heard through the dark studios and halls. **The watchman** walks through corridors and stairways from floor to floor with his swinging light cone in front of him . . . Everything is so strangely quite in this half hour from 5½ to 6. Then the watchman lets himself out through the door . . . But on the threshold he meets **the first engine man**, who hurries down to the deep cellar rooms to the thick aluminium painted pipes and the shiny containers, which make up the intestines of the Bird's Nest. The engine man turns the huge valves and steam comes streaming with a pressure of 13 atmospheres*

> *from as far away as the electricity plant . . . Right when the Bird's Nest is beginning to get the body warm, the **caretaker** arrives with **26 cleaning women** to get the morning toilette carried out. Brushes and scrubbers are moved and vacuum cleaners hum.*

Later, we hear about the porter in his little glass box by the entrance, the telephone woman, who answers requests and connects to the hundreds of ringing phones around the offices, and about the postmen arriving with their bags full to the top with letters, and which, after the director has checked them, are passed on to the handling office, from where messengers bring them around. All these functions were probably common functions in larger office-based companies at that time, but we also hear about control assistants and orchestra assistants, which provide technical and service functions specific for the core business of broadcasting.

Furthermore, it is a characteristic that all the mentioned functions are of an operational nature. The building-related functions include engine men, a caretaker and cleaning women, while the service related functions include among others, the handling office. The building technical functions were increased after moving to the much larger Radio House, and in 1951 a separate machine department was established in Radio House, and this was later supplemented with machine departments in TV City in the 1960s and in Aarhus in the 1970s.

A tactical function was established in 1964 called the administration office. This department became responsible for space management, planned maintenance and furniture procurement. It was recommended established by a secretariat from the Ministry of Finance, who around 1960 made thorough analyses of the organisation of DR with the aim to make rationalisation. The responsibilities of the administration office were extended over the years with both building-related and service related functions.

The strategic tasks with long-term planning of building capacity was until around 1970 taken care of by top management in DR and for the large building projects there were building committees established by the Danish government. Towards the end of the 1960s DR established an internal strategic planning secretariat as a staff function connected to the director general. Following a reorganisation of the Danish national policy for state buildings, the government appointed building committees were abandoned, and in connection to this an internal building coordination unit was established in DR as part of the strategic planning secretariat. The building coordination unit became responsible both for new building projects and the long-term real estate planning.

The development in DR's FM related functions until 1988 is shown in Figure B30.1. It is characterised by a strong horizontal division between the building and service related functions and a vertical division between the operational, tactical and strategic levels. The vertical division was strongest for the building-related functions, but it was also to some degree existing for the service related areas. It is at the same time characteristic that the operational functions existed much earlier as separate functions, while the tactical and strategic functions

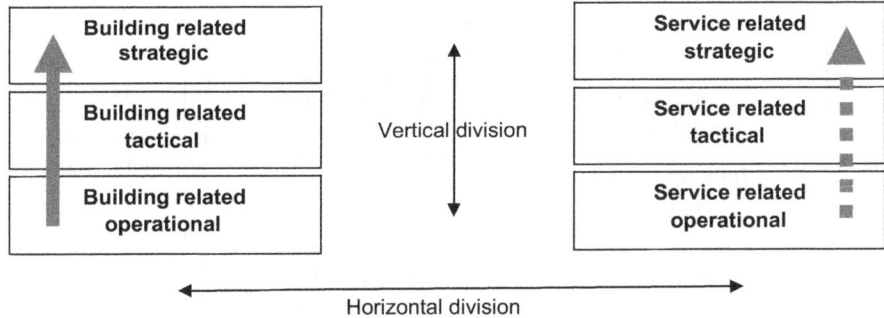

Figure B30.1 Development in the building and service related functions in DR 1925–1988

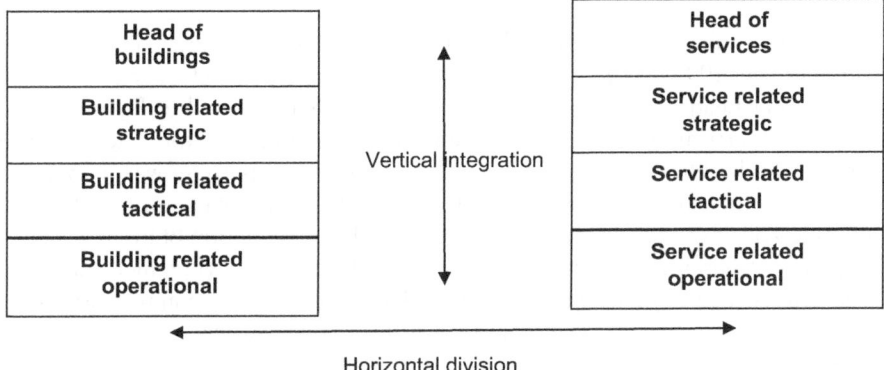

Figure B30.2 The vertical integrated organisation in DR from 1988

emerged later, along with the development of the overall organisation and its rapid expansion.

An important change in the FM related functions occurred around the time that DR lost its monopoly on national TV in 1988. This is illustrated in Figure B30.2. All these functions now came under the responsibility of the finance director, who appointed a head of buildings in charge of a building administration and a head of services in charge of a service administration. Both administrations included functions on an operational, tactical and strategic level. Thus, it involved a vertical integration, while the horizontal division between the building and service related function was kept. A primary purpose was to create a more professional organisation, which for the building-related area, among other things, involved that staff with administrative office background were substituted or supplemented with new employed staff with further educations as engineers, architectural technologists and technical assistants.

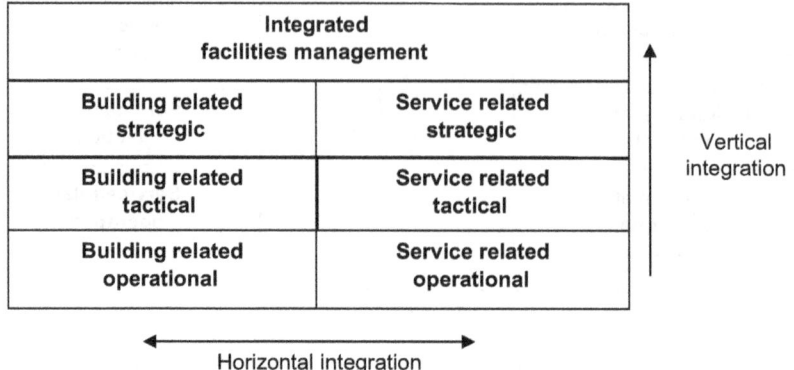

Figure B30.3 The integrated FM function in DR from 1993

The development towards a fully integrated internal FM function in DR was completed in 1993, where a new head of services was appointed and replacing the two former heads of buildings and services. This is illustrated in Figure B30.3, where the vertical integration from 1988 is supplemented by a horizontal integration of the building and service related functions. The purpose was to create a more costumer focussed FM organisation, and it was also the aim to achieve a considerable increase in efficiency, for instance by introduction of internal debiting and outsourcing of operational functions. The further development has been characterised by a number of organisational adjustments and comprehensive outsourcings, but the main principle of a fully integrated FM function has not changed.

Table B30.1 shows an overview of the development in the FM related functions in DR in the period 1928–1993. The functions firstly included the building committees, which primarily consisted of the political appointed committees in the period up to the early 1970s. Secondly, the functions included the internal building client who primarily consisted of the building coordination unit from 1970 to 1988. Thirdly, the functions included an integrated management, which only existed from 1988 after establishing the building administration and service administration. Fourthly and fifthly, they consisted of the strategic and tactical functions, and these only existed after establishing the administration office in 1964, which until 1970 were responsible for both levels, but from 1970 and until 1988 the strategic tasks were taken over by the building coordination unit, while the administration office mostly took care of tactical tasks. From 1988, both levels became the responsibility of a section for planning and execution in the building administration. Sixthly and seventhly, the operational functions were divided in technical operation and various services, and these functions were until 1988 the responsibility of the machine departments and the handling office, respectively. From 1988, they were taken over by the operational units in the building administration and by the service administration.

Table B30.1 FM related functions development in DR in the period 1928–1993

Year	1928	1940	1964	1970	1974	1988	1993
Building committee	Established by the Danish Government for Radio House and TV City						
Internal client				Building coordination			
Integrated management						Building administration	
Strategic			Administration office	Building coordination		Planning and Execution Unit	Operational Units
Tactical				Administration office			
Operational			Machine department				
Various services	Handling office, etc.					Service administration	

It is remarkable, how the functions have developed over time from only covering operational tasks towards a higher amount of the tactical and strategic tasks. In the period from 1970 to 1988, there even was a situation, where DR had a separate organisational unit on operational, tactical and strategic level for the building-related functions. The operational level (the machine departments) referred to the technical director, the tactical level (the administration office) referred to the financial director, and the strategic level (the building coordination unit) referred to the director general. The division on these task levels are shown as vertical division of FM in Figure B30.1. This was particularly strong for the building-related area, and with the establishing of the building administration in 1988 these levels were merged together in a united organisation with a common management, thus resulting in vertical integration. The horizontal division between the building oriented and the service oriented areas was kept in the whole period and reinforced in 1988 with the stabling of the service administration. I call this division the horizontal division of FM.

Table B30.2 shows a similar overview for the period 1993–2005 until the relocation to DR City. The functions building committees and internal building client here first appeared in 1999, where the organisation for DR City was established. For the other functions, the vertical united organisations are continued from the former period, but it is extended in 1993 with a unification across the horizontal division in building oriented and service oriented areas, thus resulting in horizontal integration. The principle organisation of the functions is kept, and here we can talk about an integrated FM function.

After the relocation to DR City the FM organisation was changed again and became a small, internal, but still integrated FM function, which mainly became a contract management unit following comprehensive outsourcing. However, in the latest sourcing process DR chose to insource most of the technical building

Table B30.2 FM related functions development in DR in the period 1993–2005

Year	1993	1997	1999	2005
Building Committee			Established by DR's board for DR City	
Internal Client			Client Organisation for DR City	
Integrated Management	DR Service DR Real Estate Administration Service Units Distribution, Reception, etc.	DR Service DR Real Estate Service Units Distribution, Reception, etc.	DR Internal Service DR Buildings Service Units Distribution, Reception, etc.	DR Service and Administration (SA) DR SA Real Estate DR SA Estate Operation Distribution, Reception, etc.

operation in 2016 and change from an integrated FM contract (I-FM) with one large provider to a combination of bundled and single service contracts. I have treated this development in more detail in a recent scientific article, see the literature guide. In 2017, the internal FM unit in DR was awarded the Danish annual FM prize by the Danish Facilities Management Association (DFM).

Literature guide

The chapter is mainly based on the following scientific article:

Jensen, Per Anker: The origin and constitution of facilities management as an integrated corporate function. *Facilities*, Vol. 26, No. 13/14, 2008, pp. 490–500.

A collection of research paper with a focus on DR as building client and building operator over time can be found in the following:

Jensen, Per Anker: *Space for the digital age: Defining, designing and evaluation a new world class media centre.* Research Report R-175.BYG-DTU. 2007.

The latest development in DR's sourcing and procurement of FM is treated in the following scientific article:

Jensen, Per Anker: Strategic sourcing and procurement of facilities management services. *Journal of Global Operations and Strategic Sourcing*, Vol. 10, No. 2, 2017.

Use of the model

As shown in the table that follows, the model is particularly useful in one of the five processes presented in part A: Organisational design. The model can be used as a frame of understanding and as a collection of examples of the organisational development of FM in large corporations.

Process	Phase							
Strategy development	A	B	C	D	E	F		
Organisational design	A	B	C	D	E	F		
Space planning	A	B	C	D	E	F	G	H
Building project	A	B	C	D	E	F	G	H
Process optimisation	A	B	C	D	E	F		

For organisational design, the model can be used by managers and staff on the strategic levels in FM and building client organisations in clarification of the current situation, how we should be organised in the future and our need for new knowledge and competences (phases B–D).

B.31 Organisation of FM in relation to core business

Per Anker Jensen

Introduction

The purpose of this chapter is to clarify the relationships between support functions and core business and which forms of relationships are seen as most appropriate for strategic and operational support functions. The results are based on literature studies and a case about DR conducted in 2004 as part of an MBA thesis, when I was deputy project director in the building client function for DR's new headquarters DR City (DR Byen).

The starting point for the research was theoretical reflections on the value chain in relation to FM based on the theory of Michael Porter, which since the mid-1990s has had a huge influence on management thinking, and which also formed the basis for the management model for FM in the first European FM standard (see Chapter B.15). Porter presents a generic value chain as a big arrow-shaped box divided into primary activities in the bottom half and support activities in the upper half. The pointed right end of the box has a slice marked 'Margin'.

The primary activities are the activities involved in the physical creation of the product, its sale and transfer to the customer, and follow-up, and they are presented as a sequence of typical functions in a production flow: logistics, manufacture, marketing, sales and service. The support activities are activities that support the primary activities and are presented as layers with typical corporate support functions: procurement, technology development, human resource management and corporate infrastructure. Dotted vertical lines crossing the support activities indicate that some support activities support only some of the primary activities, while others support all of them.

Porter's purpose in defining the value chain is to analyse the opportunities for companies to create competitive advantages, and both primary and support activities can assist in the creation of competitive advantages. Porter stresses that the distinction between primary and support activities does not imply a distinction between value-creating and non-value-creating activities. In the value chain theory, it is a fundamental precondition that support activities should add value to the primary activities, and this is similarly a fundamental precondition for FM.

In the following, I use the term core business for the part of a corporation responsible for the primary activities. Similarly, I use the term support functions for the part of a corporation responsible for support activities. Thus, support functions are typically providers to internal costumers responsible for the primary activities, while the primary functions are providers to external customers.

The value chain in DR

Figure B31.1 shows an illustration of the value chain in DR, which was used in the strategy work on DR's organisation in 1999–2000. The core business is marked with bold lines and types and include programme production and chief editorial. DR distribution was carried out by a number of external operators. These primary activities are shown as a sequence in the upper row of arrow-shaped boxes in Figure B31.1, while all the support functions are included in the lower arrow-shaped box.

The value chain in DR is comparable to Porter's generic value chain. There is the same division in primary activities and support activities. However, Porter's five generic primary activities cannot be identified. One of the critiques against Porter's value chain has been that it is not suitable for service companies. Even though the primary activities for DR are to deliver public service broadcasting, DR is still not a pure service company. DR can also be regarded as a production company producing television and radio programmes, which can be stored, copied and sold as physical products. These are some of the characteristics that distinguish tangible physical product from intangible services. Therefore, DR should be seen as a combined production and service company. This can be the reason why the five generic primary activities in Porter's value chain cannot be easily identified in DR's value chain.

In Figure B31.1 the support functions in DR are shown as a horizontal arrow placed in parallel with the arrows representing the core activities in the value chain. This can be seen as being in accordance with the way Porter illustrate the value chain even though Porter as mentioned place all activities of a corporation within one large horizontal arrow-shaped box with dotted vertical lines indicating that some support activities mainly are related to some of the primary activities, while others are related to several primary activities and corporate infrastructure relates to all of them.

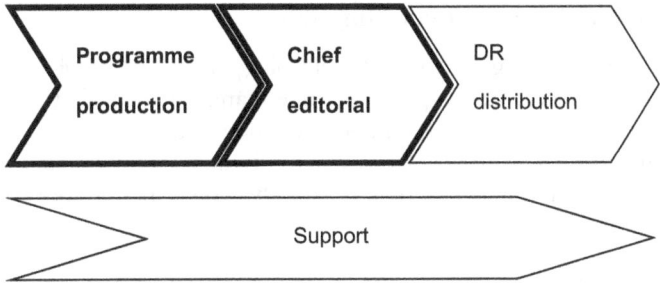

Figure B31.1 The value chain in DR

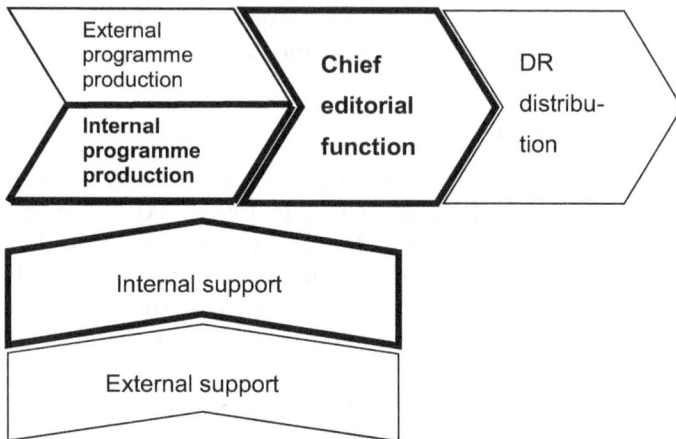

Figure B31.2 Value chains in DR including internal and external programme production and support

   Following from that, it could be relevant to define a special support value chain, which is more or less separated from the core business value chain. For the core business activities the objective is to contribute to the customer value of the end user, which for DR is the television viewers and the radio audience. For the support functions, the objective is to create value for the corporation by supporting the core business. Rather than showing the support activities as a horizontal arrow pointing towards the end users, it seems more appropriate to show the support activities as a vertical arrow pointing towards the core business activities.

   Within DR's core business, all chief editorial activities are undertaken internally, but part of the programme production has in recent years been outsourced, just like there always have external providers of foreign films, etc. The same is the case for the support activities, of course, and one of the general trends in FM is outsourcing of internal activities to outside providers. The situation with a support value chain, which include external providers, is shown in Figure B31.2.

Governance and forms of coordination

In connection to a reorganisation of technology functions in DR in 2004 it was decided to introduce a new type of resource management called IT governance involving establishment of a technology council with representatives from the different directorates. The main idea was to divide the strategic management in a number of areas and manage them separately. The principles of IT governance include the following:

•   That overall rules for the interplay between users and internal suppliers have been established

- That strategies for IT have been defined
- That fora to solve conflicts have been established
- That there is no ambiguity about who can make and change decisions

More generally, governance is a conscious management of areas of strategic importance, for instance resources like IT, personnel, finance and facilities. It is particularly relevant for larger corporations and it concerns a level of management that is above the project and departmental level and under the level of directors. This concept of governance and the idea of creating a kind of internal coalition seemed very relevant for strategic FM.

Within theory of organisational economics, forms of coordination is a central concept in analyses of organisational problems. The classical forms of coordination are market and hierarchy, which the founder of transaction cost economics Oliver Williamson calls the two basic governance structures with hybrid forms as a third alternative. The researcher Anna Grandori argues this is insufficient to cover the actual occurring variation, and she has identified seven different forms or mechanisms of coordination: price, voting, authority relationships, agent relationships, teams, negotiation, norms/customs. These are regarded as archetypes and, in reality, hybrids will often exist.

With inspiration from this, Table B31.1 present in total nine forms of coordination organised horizontally according to the degree of centralisation and vertically according to the number of decision-making parties. Between central and de-central are placed coordination forms, which either can mediate between a central and a de-central level or can be used both centrally or de-centrally. I have supplemented the seven forms defined by Grandori by the following two forms of coordination: partnership and coalition.

One-sided decision-making is equivalent to the internal line organisation with a director holding central authority at the top. He makes all important decisions, the implementation of which is often delegated to one or more semi-centrally placed management levels via agent relations down to the individual employees, who can make de-centralised decisions within a limited scope based on norms and customs.

Two-sided decision-making is the typical relationship between a customer and a supplier, where de-centralised decisions can be taken based on price alone in the simplest cases, while more complex transactions require negotiation based on

Table B31.1 Forms of coordination related to centralisation and decision-making

Decision-Making	Centralisation		
	Centralised	*Semi-centralised*	*De-centralised*
One-sided	Authority relationship	Agent relationship	Norms/customs
Two-sided	Partnership	Negotiation	Price
Multi-sided	Coalition	Voting	Team

several factors besides price and involving semi-centrally placed management at some level. A more long-term partnership or alliance relationship can be established between two parties, and this will usually involve the centrally placed top management from both parties.

Coordination of multi-sided decision-making at a de-centralised level can be done in teams working together and involving all parties. Voting is mainly related to political and other voluntary organisations and can take place at several levels, including the semi-centralised level. Multi-sided decision-making involving several organisations or sub-organisations can be coordinated by the creation of a coalition, involving more or less centrally placed representatives from each organisation. Grandori does mention coalitions in a discussion on cross-company organisations. A consortium is one type of coalition which is usually managed by a forum with a representative from each of the involved companies. Such a forum can easily run into difficulties in achieving agreement as each party has an equal power position. So a governing committee with centrally placed top managers from each party may be needed to resolve conflicts. In spite of the earlier examples, Grandori emphasises that all forms of coordination can be used both internally in a company and in cross-company collaboration.

Organisation of FM

For FM provisions with a differentiation in relation to various internal customers, de-centralised decision-making seems to be the obvious solution. That is particularly the case where the quality of the provision is easily defined and understood by both parties, and in those cases, price seems to be the best form of coordination. Examples of this could be cleaning, catering, internal removals, hiring of conference rooms and procuring of standard products. For more complex provisions with the need for dialogue about specific customisation, more centralised decision-making may be needed involving negotiation between managers at some level. Space management issues, like rebuilding projects and workplace design, could be typical examples. But for decision-making related to strategic FM, it is important to create a close collaboration and alignment between the FM organisation and the core business to achieve the necessary business orientation.

The early mentioned IT governance introduced in relation to DR's technology organisation with establishing a Technology Council can be seen as part of creating an internal coalition in DR between the technology organisation and the core business organisation with the purpose to undertake strategic resource management. In a similar way, I suggested that a Building Council should be established as part of a coalition between DR's building functions and the core business organisation with the purpose to undertake strategic resource management in relation to buildings and facilities. All main business units should be represented in the committee. Both in relation to technology and buildings, a steering group could be constituted by the corporate board of directors to supervise the collaboration and solve possible conflicts.

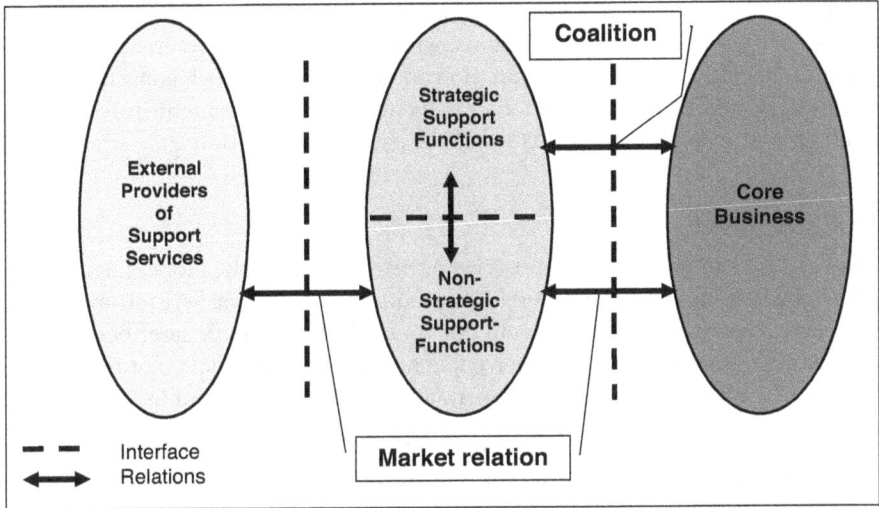

Figure B31.3 Illustration of the proposed solution

Based on this, the proposed solution to the organisation of FM in relation to core business can be illustrated as shown in Figure B31.3. A coalition is appropriate at the strategic level between core business and FM, while market relations are more appropriate both internally between operational FM and core business and between in-house FM and external providers. Establishing such a strategic coalition requires that the head of the FM function is able to talk the language of top management and make the management interested and aware of the importance of FM for the core business. A good example, where the head of FM has succeeded with that, is the case from Lego presented in Chapter B.27.

Conclusion

The analysis of FM in relation to the corporate value chain identified a particular support value chain. While the purpose of the core business is to create value in relation to external customers, the purpose of FM is to create value for internal customers by supporting the core business. FM support functions are typically carried out as a combination of internal and external provisions, and this means both internal value chains and external supply chains must be taken into account.

The relationship between core business and strategic support is characterised by a general business orientation, while the relationship between core business and non-strategic functions is characterised by a specific customer orientation. In relation to common infrastructure, it is necessary to take care of the interests of the whole corporation with a business orientation. A customer orientation towards several different internal customers will run the risk of sub-optimising in the case of common infrastructure. Therefore, a market relationship – internally or externally – is appropriate for non-strategic functions, while it is important to create a kind of coalition between a strategic FM function and the core business management. The coalition should be responsible for strategic resource management in relation to buildings. All main business units should be part of the coalition.

Literature guide

The original basis for this chapter is my MBA thesis from the Copenhagen Business School from 2004. The thesis achieved top grade. It was written in Danish and has not been published as such, but the main results have later been elaborated and published as a conference paper and a scientific article. For the article, I received a Highly Commended Paper Award 2012 from the publisher Emerald.

Jensen, Per Anker: The organisational relationships between support functions and core business. Proceedings of EFMC2007 Conference and Research Symposium: 6th EuroFM Research Symposium, Zürich: Euroforum, 2007, pp. 47–54.

Jensen, Per Anker: Organisation of facilities management in relation to core business. *Journal of Facilities Management*, Vol. 9, No. 2, 2011.

Use of the model

As shown in the table that follows, the model is particularly useful in two of the five processes presented in part A: strategy development and organisational design. The model can be used as an analysis tool, possibly in combination with the model in Chapter B.27.

Process	Phase							
Strategy development	A	B	C	D	E	F		
Organisational design	A	B	C	D	E	F		
Space planning	A	B	C	D	E	F	G	H
Building project	A	B	C	D	E	F	G	H
Process optimisation	A	B	C	D	E	F		

For strategy development, the model can be used by managers and staff on the strategic level in FM functions and their consultants for evaluation of the current strategy, defining strategy goals and developing strategy plans and as a basis for dialogue with the corporate top management (phases B–D).

*　For organisational design, the model can be used by managers and staff on the strategic level in FM functions and their consultants for evaluating the current organisation, developing proposals for new organisation and evaluation of need for new knowledge and competences, and as a basis for dialogue with internal stakeholders (phases B–D).*

B.32 The strategic FM organisation in housing associations

Per Anker Jensen

Introduction

FM in housing is different from FM in, for example, offices, production plants and private businesses. For housing, the core business is to provide housing for tenants, and it can be housing of all kinds from the lowest budget to very expensive, of different sizes, different locations and with access to different facilities. In this project, we have researched housing in multi-storey buildings in general with a focus on the importance of different forms of ownership for environmentally friendly building operation. From an environmental perspective the multi-storey housing is significant, because of the many m², and because the size of each estate provides a basis for a professional administration and building operation.

The different types of ownership give a different degree of influence to the residents and different ways of making decisions. The many small buildings in private rental, private co-ops and owner occupied dwellings mean that there are many 'small' owners and administrators in this sector, whereas in social housing, there are many relatively large housing organisations that take care of the building operation and FM for the different local boards. This chapter presents an analysis of the organisation seen from a FM point of view of the different types of ownership for multi-storey housing.

The different types of housing and ownership

Social housing (non-profit rented housing): In social housing, the residents are tenants who rent a dwelling in a social housing department, which is an independent organisational and economic unit. It is typically administered by a larger administrative social housing organisation. The residents have the right to vote at the general assembly for the housing department, who takes all the important decisions, including economy, maintenance, election of the local board, etc. This is the essence of the extensive 'residential democracy' in the sector. Social housing represents 36% of all dwellings in Danish multi-storey buildings and has a relatively high proportion of buildings between 1,000–5,000 m².

Owner occupied dwellings: These are dwellings in multi-storey buildings individually owned by the residents. Here, the common decisions concerning the building are decided by an organisation among the owners. The owner occupied dwellings represent 21% of all dwellings in Danish multi-storey buildings. The owner occupied dwellings are dominated by many small buildings (100–1,000 m²).

Private co-ops and private renting: For private co-ops, the residents buy a share of the co-op, which entitles them to rent the dwelling and to vote at the general assembly, which takes all decisions about the co-op. Over recent years, a large amount of private rented dwellings have been transformed to private co-ops, as the legislation has given the residents the possibility to buy the building when for sale. This has been very popular among the residents, who as co-op sharers gain more influence on their dwelling and building. *Private renting:* In private rented dwellings the building operation is mainly decided by the owner, and residents/tenants have limited influence. Private renting and private co-ops each represent 14% of the dwellings in Danish multi-storey buildings. They are in Denmark dominated by many small buildings (100–1,000 m²) with a limited number of dwellings.

Others: The other types of ownership in Denmark are private limited companies and public authorities. They represent 12% of the multi-storey dwellings but have been left out of this research.

Housing associations seen as FM organisations

The organisation of FM in housing has been analysed in accordance with the management model in the first European standard EN15221–1 from 2006. This involves a distinction between a demand side and a supply side with FM mediating on strategic, tactical and operational level. Table B32.1 shows the parties involved in the organisation according to these divisions for the different types of ownership of housing.

From Table B32.1 it becomes clear that the organisation for owner occupied housing and private co-ops is in principle similar. As a contrast, it is only private renting, that has FM mediation and providers represented on the strategic level. In the following, the organisation of the different types of ownerships will be analysed in more detail.

The strategic FM organisation

FM takes place in different organisational settings, as argued in the previous section. For the most advanced housing departments in practice, the various FM elements are administered by a number of different actors, with different roles and with different formal and practical possibilities for affecting FM and new FM practices. One of our conclusions in the research project is, that there is not just one FM operator; sustainable FM is practised in a network. The role of the formal FM operator can therefore be characterised as 'network management', where

Table B32.1 Organisation of FM related to type of ownership in multi-storey housing

Ownership	Level	Demand side	FM mediation	Supply side
Social housing	Strategic	National or regional housing association		
	Tactical	Local tenants' boards	Specialist staff	Consultants
	Operational	Tenants	Local inspectors	Mostly in-house staff centrally and locally
Owner occupancy and private Co-ops	Strategic	Annual association assembly		
	Tactical	Association board	Association chair	Consultants
	Operational	Residents	Association chair	Private administrator and local providers
Private renting	Strategic	Director from owner organisation	Director from private provider	Private provider and consultants
	Tactical	Manager from owner organisation	Manager from private provider	Private provider and consultants
	Operational	Tenants	Inspector from private provider	Private provider and sub-providers

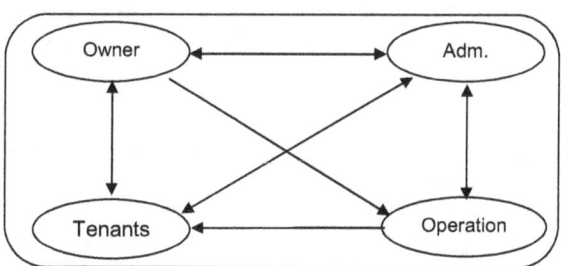

Figure B32.1 The strategic FM organisation in generic version

social and communicative skills are as important as technical expertise. We use the term 'Strategic FM Organisation' (SFMO) to name the unity of tenants, administrators, owners and operators, which all are some kind of decision maker in how to take care of housing, in practice. Figure B32.1 shows an illustration of the SFMO in a generic version.

When analysing housing associations based on the management model from the European FM standard, one challenge is that FM in housing associations is the core business, and not a support function. Or if regarded as a support function, then life and dwelling in itself becomes core business. Still the organisation has to manage its facilities and manage their demand and supply of facility services.

Table B32.1 shows that social housing associations are very self-contained organisations with most functions in-house. There is not any FM mediation and supply side on the strategic level indicated. That does not mean that such associations do not deal with strategies, but this takes place in the top management of the organisation on the demand side. In general, in relation to FM social housing associations can be said to be purely demand driven. The specialist staff indicated at tactical level of FM mediation are technical specialists in the housing association, and they mediate with consultants and providers for instance in relation to energy management and major maintenance work, but they also function as internal consultants.

Social housing is the type of ownership, which provides the most integrated frame for common decision-making as shown in Figure B32.2. The owners are the local authority and the tenants and through representative democracy, it is possible for a tenant to vote and to be elected as representatives for tenants in the social housing unit. The administration is done by the administration in the housing association and the operation is also managed within the housing association using in-house or outsourced services. The democratic rules prescribe the roles and relations between owners, tenants, administration and operation.

The owner occupancy and private co-ops resemble social housing associations by being very demand driven, but they are typically small, local organisations without many resources and very dependent on voluntary work by elected residents in their association board. Strategic decisions are solely made collectively at annual assembly meetings by all active residents. The association chair has a central role

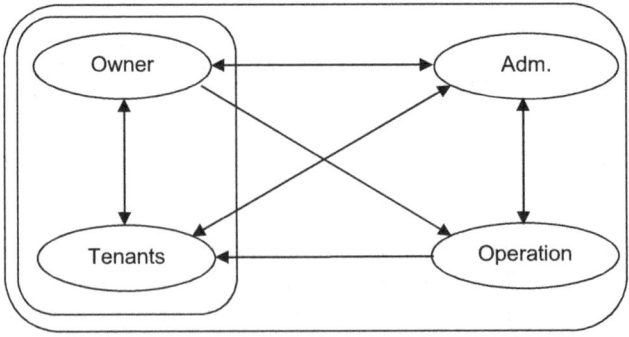

Figure B32.2 The strategic FM organisation in case of social housing

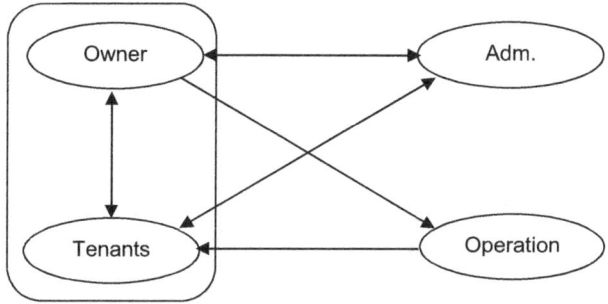

Figure B32.3 The strategic FM organisation in case of owner occupied/private co-ops

in FM mediation. The association usually has an ongoing collaboration with a private administrator, who mostly takes care of rent administration and bookkeeping. These administrators are remunerated by a fee as a percentage of the total rent and are usually selected by reputation without any economic competition. The administration is often carried out by smaller lawyer firms or similar without any technical staff.

In owner occupied/private co-ops tenants are the only owners and each housing units have its rules for representative democracy. The administration is delivered by a professional housing administrator, which sometimes is a law firm and the operation is delivered by external companies, after agreements with a board of owners, or directly by the tenants as shown in Figure B32.3.

The private renting is mostly carried out by institutional investors like capital and pension funds. There is also some private owned rented out housing. For these the situation resembles very much the one for owner occupancy and private co-ops, except that most things are managed by an owner representative and there is no association with an elected board and annual assembly. The situation indicated in Table B32.1 is related to large organisations with a private provider responsible for all facility services. This represents a quite recent development related to the general development of the FM market. The provider can be an in-house organisation, but if so, it is usually organised as a separate subsidiary company owned by the investor company and with the possibility to operate on the open market. There is an increasing trend towards economic competition between providers and towards extending the providers responsibilities for optimising the yield of the real estate investment. This increases the strategic focus on the development of the property to increase the rent and to optimise the building operation and administration. The private renting in large organisations is in this way becoming more and more supply driven.

In the case of private renting, the owner owns the real estate and leaves most often administration, contact to tenants and operation to a professional building administrator as shown in Figure B32.4. The tenants can provide the

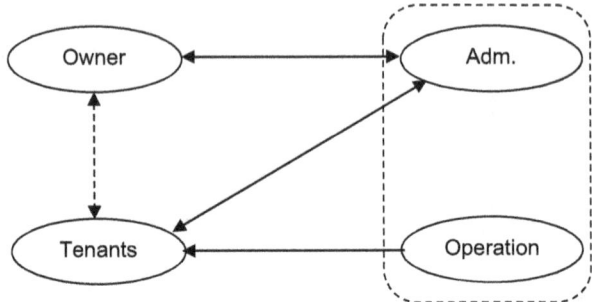

Figure B32.4 The strategic FM organisation in case of private renting

administration with viewpoints about, e.g. satisfaction levels but have no formal obligations or rights to take part in FM decision-making.

As illustrated in Figure B32.4, different types of building ownerships lead to different roles and relations between tenants, owners, administration and operation. These are important parts for the context for FM in practice, because the different stakeholders will have different roles and motivations for how they contribute to the daily use and operation of the estates as well as the FM related decision processes.

Conclusion

Different types of ownership cause, as shown earlier, different roles and relations between tenants, owners, administration and building operation in multi-storey housing. It is important to understand this as a condition for working with FM and promoting environmentally friendly and sustainable building operation in practice, because the different stakeholders will have different possibilities and motivations to participate in the daily operation of the estates and the decision processes related to FM.

Social housing is strongly demand driven with integration of demand and supply side in the same overall organisation and with a high degree of professionalism of the FM provision, which is mostly provided in-house. Owner occupancy and private co-ops are also demand driven, but the supply is not integrated and is often fragmented with division of administration and operation. Private renting is becoming increasingly supply driven by large private FM providers and the demand side is strictly divided between owners and tenants. To promote environmental building operation one needs to be aware of the different types of ownership, because it provides different frame conditions with regard to who can influence and has decision competences for everything from defining goals and initiating renovation projects to taking initiatives aimed at changing behaviour.

Literature guide

The chapter is based on the following scientific article:

Nielsen, Susanne Balslev, Per Anker Jensen and Jesper Ole Jensen: The strategic facilities management organisation in housing: Implications for sustainable facilities management. *International Journal of Facility Management*, Vol. 3, No. 1, March 2012.

The results of the research project are documented in the following research report in Danish:

Jensen, Jesper Ole, Per Anker Jensen, Morten Elle, Birgitte Hoffmann, Susanne Balslev Nielsen and Maj-Britt Quitsau: *Miljøstyret bygningsdrift i danske bolige-jendomme – under forskellige ejerformer.* SBI Report 2008:15.

The following conference paper was published based on the project:

Nielsen, Susanne Balslev, Jesper Ole Jensen and Per Anker Jensen: Delivering sustainable facilities management in Danish housing estates. *Proceedings from II International Conference on Sustainability Measurement and Modelling – ICSMM 09. Barcelona, 2009.*

Use of the model

As shown in the table that follows, the model is particularly useful in one of the five processes presented in part A: organisational design. The model can be used as a frame of understanding the development of the organisation of FM in housing associations and as a basis for research on the importance of different types of ownership for sustainable FM.

Process	Phase							
Strategy development	A	B	C	D	E	F		
Organisational design	A	B	C	D	E	F		
Space planning	A	B	C	D	E	F	G	H
Building project	A	B	C	D	E	F	G	H
Process optimisation	A	B	C	D	E	F		

For organisational design, the model can be used by managers and staff at the strategic level in housing associations to clarify the current situation, how we should be organised in the future and how we identify the need for new knowledge and competences (phases B–D).

B.33 Establishing of property centres

Susanne Balslev Nielsen

Introduction

The establishment of a central Facilities Management (FM) organisation is currently considered in many Danish municipalities to enable better coordination between the building client role and the building operator role. To learn from past experience, a research project was conceived to share the common experiences from redesigning processes of FM organisations. The 98 municipalities in Denmark are different in size but face the same challenges of ownership, construction, operation, maintenance, development and management of facilities such as schools, day-care centres, administrative buildings, sports centres, etc. The purpose of this chapter is to share experiences from centralising real estate management and hereby contribute with timely input to the municipalities' ongoing efforts to strengthen the organisation of real estate management. FM organisations in private companies can similarly learn from the municipal experience, although there may be differences in facility types and success criteria.

Six examples of FM centre formation

Six municipalities have willingly contributed to the project by sharing their experience of establishing a central real estate centre. In each municipality, the research group has spoken to the FM centre leader as well as representatives of FM employees (supply side) and a user institution (demand side). Next is a brief presentation of the individual centres, their purpose and preliminary results. This is followed by a cross-case summary of the motivation behind the centre formation, the centres' success criteria, their preliminary results and finally the experienced challenges during the transformation from a de-central organisation to a more centralised organisation.

Frederikshavn: In 2012, Frederikshavn Municipality opened a new FM centre. The goal was to reap significant economic and qualitative benefits through a coordinated and centralised approach to property issues, including freeing the economy by creating a better view of the real estate portfolio and capacity utilisation. These were incorporated into an annual savings of 1,350,000 Euro. The success criteria for the new centre were streamlining work processes, harmonising service levels and achieving savings. These success criteria have been realised through economies of scale, improved purchasing and adjustment of both resources and salaries. The focus in the next phase is shifting towards development and opportunities rather than on costs, including optimisation of operations as well as better land use and down scaling the building portfolio.

Gentofte: Gentofte Properties has existed since 1 January 2008 and is one of the oldest municipal FM centres in Denmark. The centre has passed the establishment phase and is now a 'machine at full speed'. However, this means that adjustments and modifications are still being made and is an ongoing debate. The immediate objective of the centre's formation in 2008 was to strengthen the building maintenance team and to streamline operations. These objectives have been met. Since the centre's formation in 2008, Gentofte has saved between 1,600.000–2,010.000 Euro on rationalisations and redundancies and building maintenance in the municipality is now better and more consistent.

Ishoej: Ishoej Municipality started to restructure its FM organisation in the summer of 2014. The restructuring meant, among other things, that technical services at the municipal day-care institutions were centralised under the FM centre. Similarly, the previous form of organisation was replaced by more flexible working communities that were better at completing tasks across policy areas, with a focus on tasks and customers. The aim of the restructuring was to strengthen the quality and level of service offered to the city's institutions and citizens. The restructuring has not been implemented with immediate financial savings in mind. The philosophy of the new organisation is to think in terms of total solutions and to focus on the customer.

Ringsted: In 2011, the technical service personnel in Ringsted Municipality were united in one central unit for building operation, and in 2014 this centralisation was continued with the merger of the former Municipal Properties (building client organisation) and the former Road Park into one complete municipal property centre. The primary purpose of centralisation was to achieve efficiencies, improvements and savings in operation and maintenance teams of municipal buildings. Before the FM centre was established, Ringsted Municipality had, for a long period of time, had a backlog of maintenance and found that the de-centralised maintenance funds were not always used optimally and smoothly. The centralisation should improve efficiency through better coordination and cooperation among related tasks.

Silkeborg: Silkeborg Properties opened in January 2012. This consolidated ownership of all municipal buildings and building maintenance in a central FM centre. The main incentive behind the centralisation was an overall municipal austerity in the field. The project was recognised as immediately saving approximately 400,000 Euro and has had ongoing savings of 1% annually. The primary objective of the centre has been a desire for efficiency and performance optimisation.

Svendborg: In January 2014, Svendborg Municipality opened a new FM centre, the Centre for Real Estate and Technical Service. The employees of the administration had twice before tried to centralise FM, but they could not gain political support. However, in 2013 the political support was there, and the decision was made. The centre's formation was very much focussed on achieving efficiencies and savings on property operations through economies of scale, better prioritisation and use of municipal maintenance funds. A goal was to save approximately 400,000 Euro in 2014, with further savings of 2% in 2016 and 5% in 2018 compared to the 2014 budget. This objective has so far been achieved. Another objective is efficiency through better land-use planning.

Cross-case summary: motivation, results and reflections

The six cases testify to relatively different stages of a redesign process and results, but also similarities. The reasons for motivation and success identified in the study can be summarised in the following sections:

- Good economy
- Improved building operation
- The customer in focus
- Streamlining work processes
- More robust in-house organisation
- Service to more municipal institutions (including day-care centres)
- Less backlog in building maintenance
- Coordination of related subject areas (road/park)

The results obtained by the centre formation can similarly be summarised in the following points:

- Better overview of properties and FM tasks
- Better use of the maintenance budget
- Centralised service of day-care centres
- Focus on education and competence development
- More equal service and maintenance levels within the municipality
- Team structure implemented in the FM organisation
- Cost reduction

However, there have also been a number of challenges along the way. These challenges can be summarised in the following points:

- Establishing a centre takes time and is resource demanding
- Headmasters' experience a loss of influence, because they previously have the right to dispose of facility seats, FM staff and budget
- Headmasters experience a reduced service level
- Some employees feel pressured from unsatisfied users
- Communication, involvement and dialogue are important
- Mistrust and dissatisfaction among users
- Headmasters experience that collaboration with technical services becomes more difficult

- Political and managerial support in the municipalities is crucial
- Reduced salaries and changed working conditions cause reluctance among the technical service personnel

Seven steps towards a FM centre

Based on the municipalities' experiences and recommendations, the project group developed seven steps to ease the establishment of a FM centre:

1 Start with the facilities and services, you and your co-decision makers can agree on.
2 Make a strategy for employee information and involvement.
3 Make a plan for future operation of schools/institutions early in the process.
4 Purchase expertise from the outside, if there is a lack of time or competencies in-house.
5 Make an easy contact point for the users.
6 Decide an appropriate service level for all properties.
7 Acknowledge that real estate development is a continuous development and probably never ends.

The municipalities' FM has political attention these years, because it seemed to hold an efficiency potential, but also because the building industry generally holds a potential for green innovations, and the public sector represents a niche for intelligent and sustainable products and services. The research group behind this project hope that it will be possible to attract development and research funds that can help create further professionalisation of the subject area, so real estate centres will become even more attractive jobs that efficiently and effectively deliver good FM.

Additional information about the project

The project has been completed in collaboration between CFM and the Danish union FOA organising public employees. FOA has funded the project. Their motivation was to act proactively to affect the changes experienced by technical service personnel, and CFM's motivation was to improve the knowledge base for municipal real estate management – also called municipal FM. The project report in Danish can be downloaded free of charge from www.foa.dk and www.cfm.dtu.dk

Literature guide

The chapter is mainly based on the following magazine article:

Balslev, Susanne: Centralising municipal FM organisations: Danish experiences. *EuroFM Insight*, No. 38, September 2016.

The full results of the project are documented in the following project report in Danish:

Hansen, Lars Ole Preisler and Susanne Balslev Nielsen: *Kommunal Ejendomsforvaltning – 6 fortællinger om etablering af et centralt ejendomscenter*. Research Report. CFM and FOA. 2015.

Use of the method – assessed by Per Anker Jensen

As shown in the table that follows, the method is particularly suitable for use in two of the five processes presented in part A: strategy development and organisational design. The method can be used as a process tool to guide a centralisation process. The cases can be used as inspiration to work on establishing a municipal FM centre and similar.

Process	Phase							
Strategy development	A	B	C	D	E	F		
Organisational design	A	B	C	D	E	F		
Space planning	A	B	C	D	E	F	G	H
Building project	A	B	C	D	E	F	G	H
Process optimisation	A	B	C	D	E	F		

For strategy development, the method can be used by managers and staff at the strategic level to define strategic goals and develop strategy plans and as a basis for dialogue with politicians and municipal manager (phases C–D).

For organisational design, the method can be used by managers and staff at the strategic level to develop proposals to future organisation and for evaluating the need for need knowledge and competences, and as a basis for dialogue with politicians and municipal top managers as well as the affected departments and staff (phases C–D).

B.34 Scenarios for FM in the future

Per Anker Jensen

Introduction

CFM conducted a research project in 2010–2012 on the future of FM in the Nordic countries. The purpose was mainly to identify what we should do research on in the coming years, but the purpose was also to contribute to the profession becoming qualified to take a more longsighted development perspective. Through five workshops with in total 50 participants and a questionnaire survey in Denmark, Norway, Sweden and Finland, we investigated trends in the FM sector and identified the need for new knowledge and competences. We also investigated other studies of the future of FM, which had been conducted around the same time, for instance by IFMA and the Denmark-based multinational facility service provider ISS. In this chapter, I will primarily focus on the results from the workshop in Denmark and the results of the Nordic questionnaire survey from CFM's project and compare our results with the results from ISS's study. The latter include scenarios for the future global development of FM. Finally, I will present my personal proposal for possible scenarios FM with a focus on the Nordic countries.

The starting point for CFM's project was an innovation model for the FM sector, which was developed as part of preparing the project. Four national workshops all were structured with three sessions. The first session focussed on the external megatrends, which were expected to affect the FM sector in the coming 10–15 years. The second session focussed on the internal trends and challenges for the FM sector, and the third session focussed on future need for new knowledge and competences in the FM sector. All sessions consisted of a combination of presentations, brainstorms, discussions and prioritising exercises. The fifth workshop took place during a NordicFM conference arranged by CFM in August 2011. Here the results of the four national workshops were presented and discussed with participants from across the Nordic countries. The questionnaire survey was conducted after the conference in autumn 2011 and the questionnaire was distributed to all participants in the workshops and the conference. In the questionnaire, the participants were asked to give their views on 40 statements on the future of FM divided in six themes, and the surveys got responses from 51 persons, equivalent to a response rate of 46%. Approximately half of the respondents were from Denmark.

Results from the workshop in Denmark

The workshop in Denmark was the first and took place on the 2 October 2010 at DTU with 12 participants besides two facilitators. The participants were a broad representation of professionals from private and public companies, trade organisations, and research and education institutions.

The five megatrends of the strategic environment selected as the most significant were as follows:

- Increased focus on sustainability
- Globalisation
- Mix of cultures
- Demographic change – labour shortage

Besides prioritising the different megatrends, the participants were also asked to assess, whether the individual megatrends were certain or uncertain. Both 'increased focus on sustainability' and 'globalisation' were seen as certain, while the assessment of 'mix of cultures' and 'demographic change – labour shortage' were more diverse.

Concerning trends and challenges for the FM sector, the participants were asked to make a prioritisation for the short term as well as the long term. For the short term, the following three issues were seen as the most important – and equally important:

- Sustainability: Energy, environment and branding
- Focus on price efficiency (cost savings) – not added value
- From internal (zero faults culture) to EU contracts and management of uncertainty

However, for the long term, there was a clear ranking of the three most important issues:

1   Sustainability: Energy, environment and branding
2   General agreement on the need for more education at all levels
3   Benchmarking. Structures. Standards.

In relation to the need for new knowledge and competences, the clearly most important issue was:

- Understand the client's/customer's need, particularly in development departments, and to transfer the understanding to the operational level

A number of other issues followed:

- Understand companies' organisation and their FM strategy
- Portfolio and space management

- Intelligent use of key indicators
- Sustainability
- Qualify people to make the right choices

Thus the results were a mixture of need for on one side more general and basic understanding of an ability to analyse organisations, strategies and processes, and on the other side more specific professional knowledge and competences.

Results of the Nordic questionnaire survey

Based on the results of both the literature review and the workshops 40 statements were identified as a basis for the questionnaire. The statements were clustered in the following six themes:

- Working life and style: four statements
- Resources and sustainability: eight statements
- Technology: six statements
- FM competences: five statements
- Management and new services: eight statements
- Value and professionalisation of FM: nine statements

In the survey, the respondents were among other things asked to assess each statement regarding their importance for the FM sector and which type of activities that can support the development. The results showed that 'value and professionalisation of FM' was the clearly most important theme for the FM sector followed by 'resources and sustainability'. Academic research also had most importance for 'value and professionalisation of FM' followed by 'FM competences', while academic research and industrial research and development were assessed to have more or less the same importance for 'technology'. Education and dissemination not surprisingly had most importance for 'FM competences', while public regulation mainly was of importance for 'resources and sustainability'.

All together it can be concluded that FM research in the future can have the highest importance in relation to 'value and professionalisation' and 'resources and sustainability'. Based on the statements with highest priority these themes for instance can include the focus areas mentioned next.

Value and professionalisation

- Introduction of methodologies for FM to become a critical, strategic management tool, which couples the role of facilities to the strategy of organisation's core business
- Introduction of a set of principles to measure and document the added value of FM services

Resources and sustainability

- Sustainability as a basic requirement in FM services across companies
- Introduction of methodologies to manage energy savings in FM services

ISS 2020 Vision

The main objectives of the ISS 2020 Vision study were to develop a set of global scenarios for the future of the FM and services industry and to bring awareness about the future trends, uncertainties and opportunities that could have the greatest impact on ISS and its customers.

The study was carried out by the Copenhagen Institute for Futures Studies in collaboration with GlobalFM and IFMA Foundation during the winter and spring of 2010–2011 using the following methods:

- Workshops with ISS executive group
- Surveys with responses from 308 ISS Top Managers around the world and from 50 external global facility management experts conducted in January 2011
- In-depth interviews with six external industry spokesmen
- Evaluation of the results at a customer workshop in London in July 2011

The study identifies ten megatrends divided in three groups of external trends and a group with two industry specific trends shown in Table B34.1.

Compared to the megatrends identified in CFM's study, 'sustainability' is a common megatrend. The megatrend 'globalisation' is also common for both CFM's and ISS's study. The theme 'value and professionalisation' from CFM's study is not directly included as a megatrend in ISS's study, but the theme is strongly emphasised both in the referred interviews with industry spokesmen and in the actual report text. Besides that, there are a number of trends, which are different in the two studies for Denmark and the global development, respectively.

Table B34.1 Trends identified in the ISS-study

Type of megatrend	Megatrends
Factor	1 Economic growth 2 Globalisation 3 Aging and urbanisation 4 Sustainability
Knowledge	5 Technological development 6 The growth of a knowledge society
Social	7 Individualisation 8 Commercialisation
Industry specific	9 New ways of working 10 Preparedness and populations at risk in densely populated urban areas

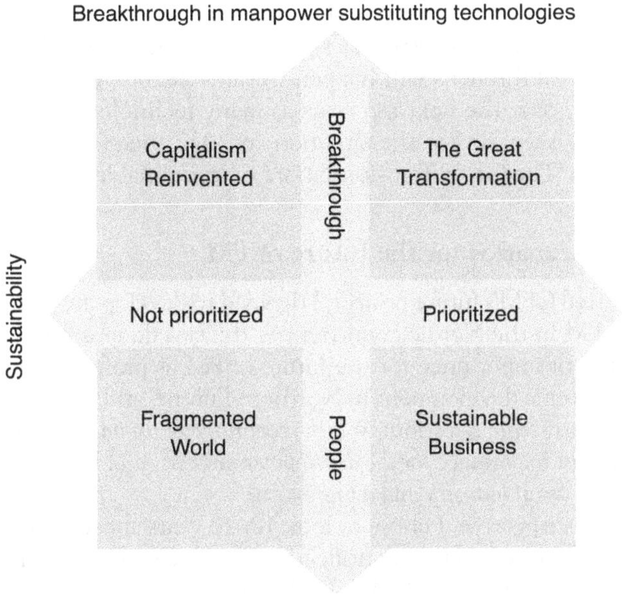

Breakthrough in manpower substituting technologies

Figure B34.1 Scenarios from ISS 2020 Vision

Based on the megatrend analysis, the study identified 'the technology dimension' and 'the sustainability dimension' as the key uncertainties, which were regarded to have the highest impact on the global FM and services industry towards 2020. This resulted in establishing the four scenarios shown in Figure B34.1.

The technology dimension represents the vertical axis divided according to the degree of breakthrough in manpower substituting technology with high at the top and low at the bottom. The sustainability dimension represents the horizontal axis divided according to the degree of sustainability being prioritised with high on the right side and low on the left side.

The four scenarios are as follows:

- *Capitalism Reinvented*, where a high degree of technology breakthrough is realised, while sustainability is not prioritised
- *The Great Transformation*, where a high degree of technology breakthrough is realised, while sustainability is also prioritised
- *Sustainable Business*, where only a low degree of technology breakthrough is realised, while sustainability is prioritised
- *Fragmented World*, where only a low degree of technology breakthrough is realised, while sustainability is not prioritised

At the customer workshop in July 2011, the participants – by and large – agreed that 'Fragmented World' best represents the world in 2020. The participants all believed that sustainability would increase in importance towards 2020 but only grow moderately. The pace of technological development and its impact on the global FM and services industry was seen as very uncertain. Most participants believed that breakthroughs will happen towards 2020, but they will not be as ground breaking over the next ten years as many technology enthusiasts opine. The most likely scenario for the situation in 2020 therefore is seen, either as being either the 'The Great Transformation' or 'Sustainable Business'.

My personal scenarios for the future of FM

When we started CFM's future project, I decided to develop my own scenarios for the future of FM in the Nordic countries. As the two dimensions with the highest uncertainty of importance for the future of FM as profession and industry, I chose 'the economic development in Northern Europe' and 'the global climate'. I formulated the strategic question: *'Which scenarios can we imagine for the conditions that has the highest importance for the development in FM, and how will they influence the future of FM as an industry and a profession?'*

As the time perspective, I chose to look 10–20 years ahead or more specific to the year 2025. This led to the identification the five scenarios name with road-metaphors as shown in Figure B34.2. In the upper part of the figure, it is at each end of the two axes described with bullet points, what characterises the positive and negative extremes for the two dimensions? The name and place of the five scenarios are also indicated in the upper part, and in the bottom part, each scenario is described with bullet points.

Conclusion

When I read the ISS-study, I was surprised how similar their scenarios are to mine. The vertical dimensions ('breakthrough in manpower substituting technology' and 'economic development in Northern Europa') are quite different, but the horizontal dimensions 'sustainability' and 'global climate' are pretty close to each other.

At the customer workshop as part of the ISS-study it was assessed that the current situation mostly resembles the scenario 'fragmented world'. I see this as the most negative scenario in the ISS-study equivalent to my scenario 'The Road on the Edge'. This assessment from the customer workshop can be seen as expression of the future hardly can be worse than today. Contrarily, I saw it as an expression of a too optimistic assessment. I feared that the global development was in risk of becoming even a lot worse. My assessment therefore was, that the present situation is closer to the middle of both Figure B34.1 and Figure B34.2, thus similar to the scenario I called 'The Roundabout'. However, it is difficult to make predictions – especially about the future!

The aforementioned were my evaluations in 2012, when I first presented the scenarios. In the introduction to Part B, I have updated my evaluation to the present situation in 2018.

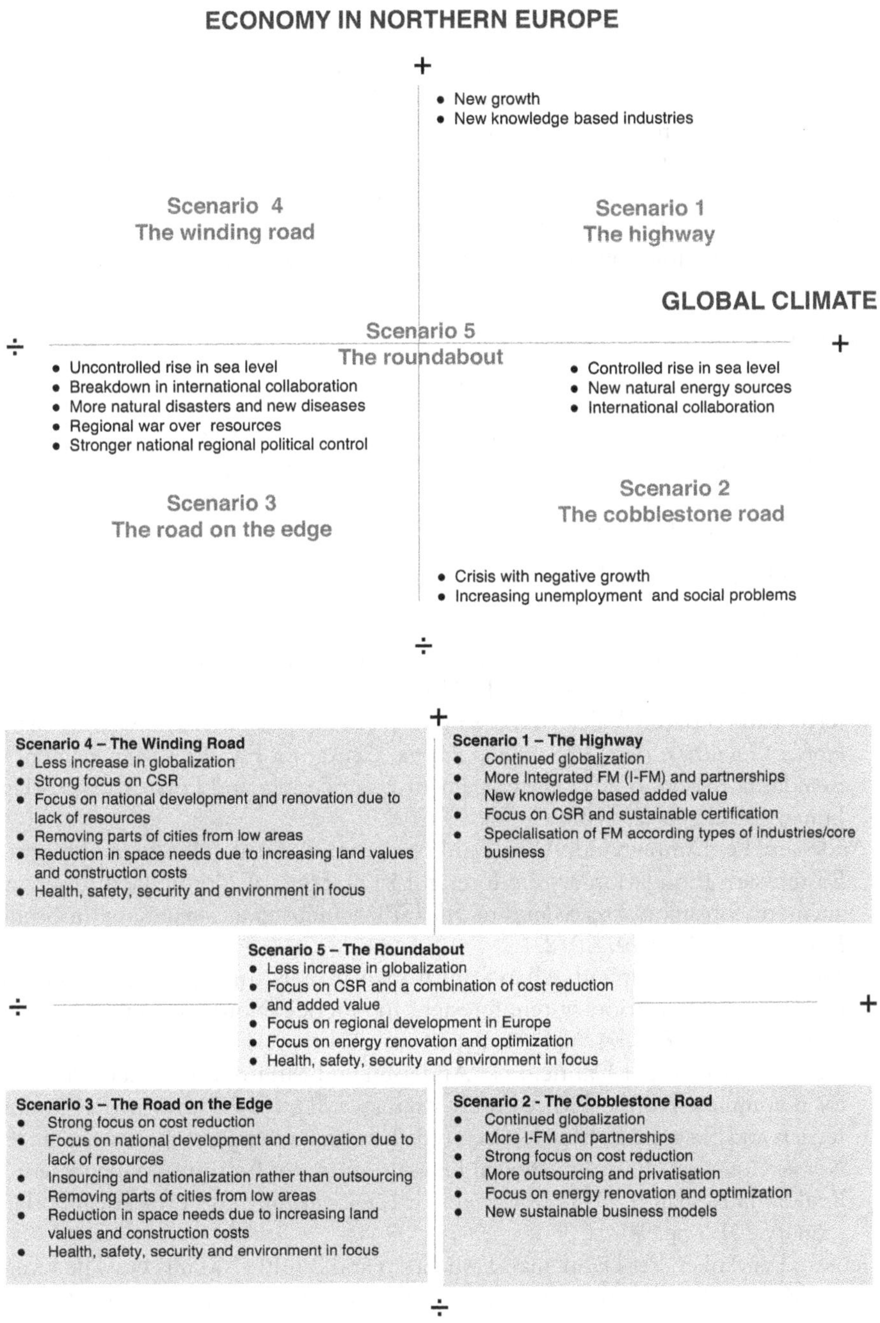

ECONOMY IN NORTHERN EUROPE

+

- New growth
- New knowledge based industries

Scenario 4
The winding road

Scenario 1
The highway

GLOBAL CLIMATE

÷

Scenario 5
The roundabout

+

- Uncontrolled rise in sea level
- Breakdown in international collaboration
- More natural disasters and new diseases
- Regional war over resources
- Stronger national regional political control

- Controlled rise in sea level
- New natural energy sources
- International collaboration

Scenario 3
The road on the edge

Scenario 2
The cobblestone road

- Crisis with negative growth
- Increasing unemployment and social problems

÷

+

Scenario 4 – The Winding Road
- Less increase in globalization
- Strong focus on CSR
- Focus on national development and renovation due to lack of resources
- Removing parts of cities from low areas
- Reduction in space needs due to increasing land values and construction costs
- Health, safety, security and environment in focus

Scenario 1 – The Highway
- Continued globalization
- More Integrated FM (I-FM) and partnerships
- New knowledge based added value
- Focus on CSR and sustainable certification
- Specialisation of FM according types of industries/core business

Scenario 5 – The Roundabout
- Less increase in globalization
- Focus on CSR and a combination of cost reduction
- and added value
- Focus on regional development in Europe
- Focus on energy renovation and optimization
- Health, safety, security and environment in focus

÷

+

Scenario 3 – The Road on the Edge
- Strong focus on cost reduction
- Focus on national development and renovation due to lack of resources
- Insourcing and nationalization rather than outsourcing
- Removing parts of cities from low areas
- Reduction in space needs due to increasing land values and construction costs
- Health, safety, security and environment in focus

Scenario 2 - The Cobblestone Road
- Continued globalization
- More I-FM and partnerships
- Strong focus on cost reduction
- More outsourcing and privatisation
- Focus on energy renovation and optimization
- New sustainable business models

÷

Figure B34.2 My personal scenarios for the future of FM

Supplementary information about the project

CFM's project was carried out with me as project manager in collaboration with my colleagues: Professor Per Dannemand Andersen and senior researcher Birgitte Rasmussen, both from the division Technology and Innovation Management at DTU Management Engineering. They together were responsible for facilitating all workshops. The workshops were arranged by me in collaboration with DFM in Denmark and other member associations of NordicFM in Norway, Sweden and Finland. Further information on the project can be found at www.cfm.dtu.dk.

Literature guide

The chapter is mainly based on the following research report:

Rasmussen, Birgitte, Per Dannemand Andersen and Per Anker Jensen: *Foresight on facilities management in the Nordic countries*. Research Report. Centre for Facilities Management – Realdania Research, DTU Management Engineering, Report 2:2012, February 2012.

The results were also published in several book chapters, conference papers and scientific articles:

Andersen, Per Dannemand, Birgitte Rasmussen and Per Anker Jensen: Future trends and challenges for FM in the Nordic countries. Chapter 9.2 in Per Anker Jensen and Susanne Balslev Nielsen (eds.): *Facilities Management Research in the Nordic Countries: Past, Present and Future*. Centre for Facilities Management – Realdania Research, DTU Management Engineering, and Polyteknisk Forlag, January 2012, pp. 10–16.

Andersen, Per Dannemand, Allan Dahl Andersen, Per Anker Jensen and Birgitte Rasmussen: Innovation-system foresight in practice: A Nordic facilities management foresight. *Proceedings of 5th ISPIM Innovation Symposium in Seoul, Korea, 9–12 December 2012*.

Andersen, Per Dannemand, Allan Dahl Andersen, Per Anker Jensen and Birgitte Rasmussen: Innovation-system foresight in practice: Nordic facilities management foresight. *Futures*, Vol. 61, September 2014, pp. 33–44.

Jensen, Per Anker, Per Dannemand Andersen and Birgitte Rasmussen: Proposal for a common Nordic FM research strategy. Chapter 9.5 in 2 in Per Anker Jensen and Susanne Balslev Nielsen (eds.): *Facilities Management Research in the Nordic Countries: Past, Present and Future*. Centre for Facilities Management – Realdania Research, DTU Management Engineering, and Polyteknisk Forlag, January 2012, pp. 10–16.

Jensen, Per Anker, Per Dannemand Andersen and Birgitte Rasmussen: The future of FM in the Nordic countries and a possible common research agenda. *Proceedings of the 11th EuroFM Research Symposium, 24–25 May in Copenhagen, Denmark*. Centre for Facilities Management – Realdania Research, DTU Management Engineering, and Polyteknisk Forlag, May 2012.

Jensen, Per Anker, Birgitte Rasmussen and Per Dannemand Andersen: Research on the future of FM in the Nordic countries. *Proceedings of IFMA World Workplace Conference, in San Antonio, USA, 31 October–2 November 2012.*

Jensen, Per Anker, Per Dannemand Andersen and Birgitte Rasmussen: Future research agenda for FM in the Nordic countries of Europe. *Facilities*, Special Issue, Vol. 32, No. 1/2, 2014.

My personal scenarios for the future of FM have been presented verbally at FM conferences in Denmark and the US, but they have not been published before in English.

Use of the tool

As shown in the table that follows, the model is particularly useful in one of the five processes presented in part A: strategy development. The model can be used as an analysis tool, and the described scenarios can be used as inspiration.

Process	Phase							
Strategy development	A	B	C	D	E	F		
Organisational design	A	B	C	D	E	F		
Space planning	A	B	C	D	E	F	G	H
Building project	A	B	C	D	E	F	G	H
Process optimisation	A	B	C	D	E	F		

For strategy development, the model can be used by managers and staff on the strategic level in FM functions and their consultants to evaluate their current strategy and to define new strategy goals (phases B–C).

Index